A CENTENNIAL HISTORY
EASTERN BUILDING MATERIAL DEALERS ASSOCIATION
1892–1992

By George W. Franz

Penn State Delaware County

Presented by John & Mary Rearick.
J.H. Rearick & Son, Inc
Page 40 + 102 + 144

A CENTENNIAL HISTORY
EASTERN BUILDING MATERIAL DEALERS ASSOCIATION
1892–1992

By George W. Franz
Penn State Delaware County

THE
DONNING COMPANY
PUBLISHERS

Sawyers bucking (cutting) felled trees into log lengths with foreman looking on. Team of bark peelers are at work in the background. Circa 1890–1910. Photo courtesy of Pennsylvania Lumber Museum

> "Remember the Past,
>
> Honor the Present,
>
> Look Forward to the Future"
>
> Pennsylvania Lumbermen's Protective Association (1892–1903)
>
> Pennsylvania Lumberman's Association (1903–1922)
>
> Pennsylvania Lumbermens Association (1922–1933)
>
> Middle Atlantic Lumbermens Association (1933–1984)
>
> Eastern Building Material Dealers Association (1984–1992)

Copyright © 1992 by Eastern Building Material Dealers Association
All rights reserved, including the right to reproduce this work in any form whatsoever without permission in writing from the publisher, except for brief passages in connection with a review. For information, write:

The Donning Company/Publishers
184 Business Park Drive, Suite 106
Virginia Beach, Virginia 23462

Richard A. Horwege, Editor
Eliza Midgett, Designer
Debra Y. Quesnal, Project Director
Holly B. Nuechterlein, Project Research Coordinator
Elizabeth B. Bobbitt, Pictorial Coordinator

Library of Congress Cataloging in Publication Data:
Franz, George Wiliam, 1942
  A centennial history : Eastern Building Material Dealers Association, 1892–1992 / George W. Franz.
    p. cm.
   Includes index.
   ISBN 0-89865-831-4
   1. Eastern Building Material Dealers Association—History. 2. Building materials industry—United States—Societies, etc.—History. 3. Lumber trade—United States—Societies, etc.—History. I. Title.
HD9715.8.U63A1214   1992   91-38500
381'.4568—dc20   CIP
Printed in the United States of America

Contents

Foreword by B. Harold Smick, Jr., Chairman Centennial Committee — 7

Introductions — 10

Chapter I Introduction: Of Lumbermen and Trade Associations — 15

Chapter II The Early Years — 39

Chapter III The Middle Years: Middle Atlantic Lumbermens Association — 71

Chapter IV Eastern Building Material Dealers Association Today — 105

Chapter V The Annual Meeting — 119

Chapter VI Insurance — 129

Chapter VII Publications — 135

Then and Now — 142

Chief Staff Officers — 150

Association Offices — 151

Centennial Companies — 154

Elected Leadership 1991 — 155

Past Presidents — 156

Gavel Cavaliers — 157

Active Dealer Members — 158

Associate Members — 163

Committees — 165

EBMDA Staff — 173

Business Profiles — 177

Index — 186

About the Author — 192

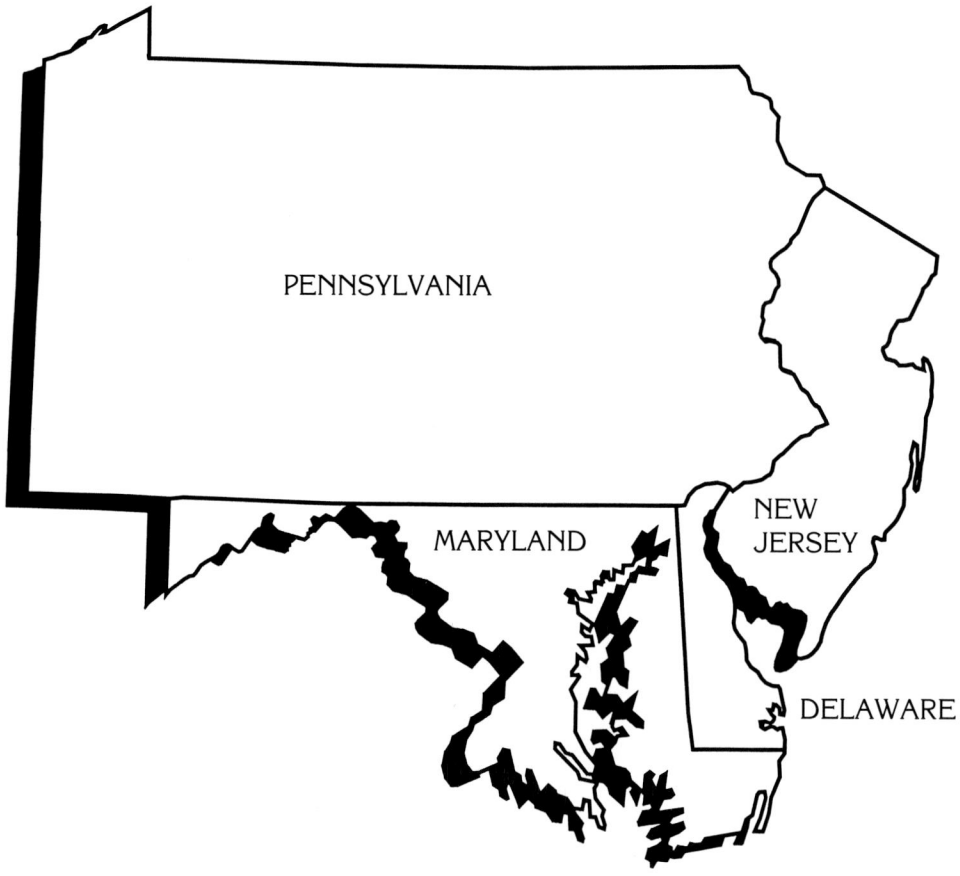
EBMDA Service Area

Foreword

*"Remember the Past,
Honor the Present,
Look Forward to the Future"*

From a very small beginning in Reading, Pennsylvania, on March 22, 1892, the Pennsylvania Lumbermen's Protective Association has emerged one hundred years later as the Eastern Building Material Dealers Association, one of the giants among the twenty-three federated groups serving the retail lumber industry and building materials in the United States.

Your Centennial Committee proudly presents a history of our first one hundred years of industry service. We are fortunate to have had as our historian and author, Dr. George W. Franz, honors coordinator, Penn State Delaware County Campus. By his painstaking and careful research, Dr. Franz has prepared this centennial issue. Without his dedicated efforts, this book would not have been possible.

We are most grateful to our very capable EBMDA staff at the Media, Pennsylvania office for their cooperation, research and special efforts. Special mention is due our executive vice president, David B. Kreidler, and senior vice president and secretary, Harry H. Johnson III, for their continuous support.

Your Centennial Committee expresses appreciation to the board of EBMDA for their support and encouragement in making this centennial issue possible.

*B. Harold Smick, Jr.,
Chairman Centennial Committee
Smick Lumber & Building Materials,
Quinton, N.J.
Photo courtesy of Smick Lumber and Building Materials*

We are especially proud of the members and associate members who have remained in continuous operation for more than one hundred years. Congratulations to each of you.

We acknowledge with gratitude the many lumber dealers who shared their historic records, photos, and memorabilia with us. Many of these are included in this book.

It has been no easy task to write the history of Eastern Building Material Dealers Association as not all the years' records were available. Written particularly to fill the need for an authentic history of our trade association, you will find articles, some old and some new, on persons, places, things, and events during our first one hundred years of leadership.

It is our hope you will find this commemorative journal an interesting history of your association's first century, our involvement in the retail lumber industry, some of the people and its leaders who did their part toward making our association and our industry so outstanding.

B. Harold Smick, Jr.
Chairman, Centennial Committee

1991-92 Centennial Committee

Honorary Chairmen:
G. Hunter Bowers
Joseph Brosius (deceased)
Watson Malone (deceased)
John D. Mitchell (deceased)

Committee Members:
Mr. B. Harold Smick Chairman Smick, Lumber & Building Materials, Quinton, N.J.
Mr. John H. Auld, John H. Auld & Bro. Co., Allison Park, Pa
Mr. David Birchmire, Deepwater Lumber & Supply, Pennsville, N.J.
Mr. Robert M. Bushey, Cavetown Planning Mill, Cavetown, Md.
Mr. Frank Braceland, Villanova, Pa.
Mr. Charles C. Cluss, O. C. Cluss Lumber Co., Uniontown, Pa.
Mr. Leonard Desmet, Desmet Lumber & Supply Co., Cecil, Pa.
Mr. John W. Eckman, Eckman Lumber Co., Inc., Lehighton, Pa.
Mr. Arthur R. Borden, Lewisburg Builders Supply, Lewisburg, Pa.
Mr. Farrell L. Goble, Brosius-Eliason Company, New Castle, Del.
Mr. Frank M. Hankins, Jr., H. H. Hankins & Bro., Bridgeton, N.J.
Mr. Lee R. Harman, U. L. Harman, Inc., Marydel, Del.
Mrs. Annabelle Hornsby, Allied Building Center, Salisbury, Md.
Mr. James J. Maloney, Jr., Sykes Scholtz Collins Lumber, Philadelphia, Pa.
Ms. Alta M. Miller, Peachey Builders, Belleville, Pa.
Mrs. Mary K. Rearick, J. H. Rearick & Sons, Inc., Dillsburg, Pa.
Mr. Thomas F. Rider, Faxon Lumber Company, Williamsport, Pa.
Mr. J. Fred Robinson, Newark Lumber Company, Newark, Del.
Mr. A. K. Shearer III, A. K. Shearer Company, North Wales, Pa.
Mr. William E. Shone, Jr., Shone Lumber & Building, Stanton, Del.
Mr. John E. Smith, Jr., Smith Building Supply, Churchton, Md.
Mr. William A. Smith, Jr., Middlesex Building & Roofing Supply, Middlesex, Del.
Mr. Vincent J. Tague, Tague Lumber, Inc., Philadelphia, Pa.
Mr. Claude S. Wetherill, 3rd, Wetherill Lumber, Inc., Bristol, Pa.
*Mr. James P. Rauch, Crafton Lumber & Supply, Pittsburgh, Pa.
*Mr. Bruce C. Ferretti, Lehigh Lumber Company, Bethlehem, Pa.
*David B. Kreidler
*Harry H. Johnson III

*Ex-officio

Bruce Ferretti, Chairman, Eastern Building Material Dealers Association Photo courtesy of Lehigh Lumber Company

EASTERN
Building Material Dealers Association

One hundred years. It's hard to believe the Eastern Building Material Dealers Association has been in existence for one hundred years. Beginning as the Pennsylvania Lumbermans Protective Association in 1892 and progressing through the Middle Atlantic Lumbermans Association in 1933, today we are an organization of 800 members, branches and associate members in 4 states and the District of Columbia.

This Centennial Journal is dedicated to all the volunteers who have made this Association the finest in the United States. Our Health and Casualty Insurance programs are outstanding. Our retirement plan has over 37 million dollars invested. The Educational Foundation puts a high priority on education for dealer members, plus our many other member services insure independent lumber and building material dealers will survive and prosper for another 100 years. Our Association Staff is both dedicated and professional.

I wonder what the Chairman 100 years from now will have to say?

BRUCE FERRETTI, Chairman
Eastern Building Material
   Dealers Association
May, 1992

604 East Baltimore Pike • Media, PA 19063 • Telephone 215/565-6144
Fax# (215) 565-0968

NATIONAL LUMBER AND BUILDING MATERIAL DEALERS ASSOCIATION

January 1, 1992

To The Members Of The Eastern
Building Material Dealers Association

It is a great pleasure to send hearty congratulations and warmest best wishes to the Eastern Building Material Dealers Association as it completes its first century of service.

EBMDA's members have made immense contributions to their communities since its founding in Wilkes-Barre in 1892. We dealers, across the country, have united to learn from each other and work together to improve the legislative and economic environment where we operate. EBMDA has been especially effective representing its members in a variety of challenging legislative forums, creating solid learning opportunities at its shows, conventions and other meetings, and using the mass buying power of its members to help deliver economy and efficiency to all.

Again, congratulations from your National Association, its officers, directors and members. EBMDA's creative, dedicated and tenacious leadership has it well prepared to begin its second hundred years. We are proud of your accomplishments and service, and grateful for your support and friendship.

Sincerely,

J. Howard Luck
NLBMDA President

40 IVY STREET, SE • WASHINGTON, DC 20003 • (202) 547-2230
Facsimile 1-202-547-7640

J. Howard Luck, President, NLBMDA
Photo courtesy of NLBMDA

R. William Taylor, President, American Society of Association Executives
Photo courtesy of American Society of Association Executives

AMERICAN SOCIETY OF ASSOCIATION EXECUTIVES
The ASAE Building
1575 Eye Street, NW
Washington, DC 20005
202-626-2700

R. William Taylor, CAE
President

March 20, 1991

David B. Kreidler, CAE
Executive Vice President
Eastern Building Material Dealers
 Association
604 Baltimore Pike
Media, PA 19063

Dear Dave:

My sincere congratulations to the Eastern Building Material Dealers Association on reaching your Centennial. As one of the oldest associations in this country, you should be commended for serving and prospering for 100 years.

Your association has helped its members in good times and in lean times by promoting business for the industry, encouraging ethical practices among your members, setting industry standards, providing educational offerings that keep your members informed, and monitoring legislative issues.

Business in Ameria would be quite different without the significant impact of associations like yours. For instance, associations spend $14.5 billion annually to set and maintain product and safety standards that protect all consumers.

Even the government relies on the extensive statistical and research programs conducted by associations -- from business statistics that enable both government and the business community to forecast and plan better for economic trends to research in a variety of technical areas.

To that end, 90 percent of all associations offer information to their members in the form of educational seminars. In 1989 associations spend more on education than every state except California. This education helps members keep abreast of the competition as well as new developments in the industry, and provides them with much needed technical expertise and management skills.

The Eastern Building Material Dealers Association should be proud of its history, and for its part in the bigger picture, proving once again that associations truly do advance America.

Cordially,

R. William Taylor

RWT/wmb

EASTERN
Building Material Dealers Association

The building material dealer is a breed apart. . . .a symbol of an American success story. . . .usually in the setting of a family-owned business.

Of the tens of thousands of business types in the United States, only a very few can claim the fraternal instincts one so often sees in the retail building material industry. It is safe to say that any retail building material dealer could travel to almost any location in the United States and instantly strike up an acquaintanceship/ friendship with another dealer. Locally, a remarkable networking process helps shape the image of retailers, wholesalers, manufacturers, and co-ops alike. Broadly speaking, an adherence to ethical standards survives despite growing competitiveness.

Often the building material retailer occupies the enviable spot of "pillar" in the community. Considering the values associated with real estate and inventory, the dealer is almost always an individual of considerable substance.

In our combined half-century of service to this industry, we have been filled with pride at the prospect of representing your interests and in some way, helping you all to progress as a family, as a business, as an association and as a community. As this nation grows and prospers, so shall this industry endure for generations to come. We salute all of you for your contributions to Eastern Building Material Dealers Association, for your integrity and honesty and for your foresight in planning for the future.

David B. Kreidler, CAE
Executive Vice President

Harry H. Johnson III, CAE
Senior Vice President

Seated: Harry H. Johnson III, CAE - Senior Vice President/Secretary.
Standing: David B. Kreidler, CAE - Executive Vice President
Photo courtesy of EBMDA

February 12, 1992

David B. Kreidler, Executive Vice President
Harry H. Johnson III, Senior Vice President/Secretary
Bruce Ferretti, Chairman
B. Harold Smick, Jr., Chairman Centennial Committee
Dr. George W. Franz, Author
Eastern Building Materials Dealers Association
604 East Baltimore Pike
Media, Pennsylvania 19063

Dear Mr. Kreidler, Mr. Johnson, Mr. Ferretti, Mr. Smick, Dr. Franz, and Members of Eastern Building Materials Dealers Association:

Congratulations on the One Hundredth Anniversary of your organization. This handsome and fascinating volume, "A Centennial History, Eastern Building Materials Dealers Association, 1892 - 1992" is an elegant testimony to the hard work and spirit of entrepreneurship which have envigorated EBMDA for the last century.

Unfortunately, even with careful attention to the details of all steps of production, there is a significant error on page 116. The chart "Mid-Atlantic Risk Management - Premium" was omitted and another chart wrongly printed in its place. This is entirely the fault of Donning Company/Publishers. A corrected copy of page 116 is provided which can be slipped into the book at the correct place. We apologize for this error and the inconvenience it causes.

Again, congratulations on the achievements of your first one hundred years. We hope you will thoroughly enjoy this grand book. This is a volume which we will proudly show to other corporations considering a similar project.

Sincerely,

Steve Mull, General Manager
The Donning Company/Publishers

ST:bb

Chapter 1
Introduction: Of Lumbermen and Trade Associations

LUMBERING IN PENNSYLVANIA

The lumber industry and related activities is one of the oldest enterprises in America. The earliest explorers to North America came in search of gold and found trees and some of the earliest exports were lumber products. As one would expect in a state whose name means "Penn's woods," lumbering was one of the first activities in colonial Pennsylvania, and this accounted for a major portion of the economic growth and well being of the commonwealth well into the twentieth century.

From the very beginning, William Penn realized the importance of the forests to Pennsylvania's well being and he decreed in 1681 that one acre in five had to remain forested. From the opening of the first sawmill in 1662, on what would eventually be called the Delaware River, to the beginning of the twentieth century, Pennsylvania was always among the top ten states in the number of sawmills and the quantity of lumber produced. It also ranked first in the number of mills from 1860 to 1900 and in the top four in lumber production during that time. After the turn of the century, lumbering in Pennsylvania declined until, by 1940, Pennsyl-

Notching to direct the fall of the tree. Tools include a bark spud, double-bit axe, crosscut saw. Note kerosene jug on ground at right, used to clean pitch from the saw. Photo courtesy of Pennsylvania Lumber Museum

ranked twenty-third among the states in the number of sawmills and lumber produced. Consequently, the lumber industry and wood-using industries within the Commonwealth were a major voice in its social, economic, and political life.

As a result, Pennsylvania passed some of the earliest forest conservation laws and, as we will see later in this work, the Pennsylvania Lumbermen's Protective Association would support these efforts, which would place Pennsylvania, by the beginning of the twentieth century, in the front ranks of states preserving and caring for its forests.

This rich, varied, and powerful lumber industry was both a blessing and bane to the retail lumber dealer in Pennsylvania. On the one hand, lumber was easily and readily available at relatively cheap cost because of the low transportation costs. However, because of the pervasive nature of the lumber industry throughout the state, it was easy for the consumer to go directly to the manufacturer or wholesaler for his lumber. As a result, it became increasingly important for the retail lumber dealers to develop some mechanism to curtail this threat to their markets.

TRADE ASSOCIATIONS

Trade associations have been organized in America since the middle of the nineteenth century. While their creation was sporadic around the time of the Civil War, there was a noticeable acceleration in their formation with the onslaught of the economic crises of the 1870s and 1890s. These early efforts had, as their primary focus, agreement on prices among competitors and the regulation of production.

These efforts were nothing new, and similar efforts go back to the beginning of recorded time. Adam Smith commented on these kind of activities in his 1776 work, *The Wealth of Nations*, when he wrote: "People of the same trade seldom meet together, even for merriment and diversion but the conversation ends in a conspiracy against the public or in some contrivance to raise prices."

Within the English tradition, price fixing and the curtailment of production were prohibited by the common law. As a result, "gentlemen's agreements" were worked out and, when these type of arrangements proved ephemeral, trade associations developed as another mechanism to attempt to regulate price and production among competitors.

With the rapid industrialization of the United States in the late nineteenth century, most industries developed some type of trade association in an effort to provide some stability and regularity to the marketplace. However, while business and industry were moving in one direction, public policy was moving in another. The Sherman Act of 1890, which outlawed any contract, combination, or conspiracy in restraint of trade, committed the country to competition as an economic policy. It was the first of a series of antitrust legislation at the national level to attempt to insure competition as national policy.

The Supreme Court of the United States was of two minds in interpreting the Sherman Act and other pieces of antitrust legislation as it related to business activities. In *United States v. Eastern States Retail Lumber Dealers' Association*, 234 U.S. 600 (1914), the court declared "illegal *per se*" any activity in the form of control of output, market sharing, and the exclusion of competitors by boycott or other means or other efforts to influence price in interstate commerce.

Typical wood scene. Logs will probably be moved down a slide to a landing. Photo courtesy of Pennsylvania Lumber Museum

At the same time, however, the court also held to the doctrine of "reasonable restraint" which allowed for the expansion of trade associations as one way to conform to the antitrust legislation in the United States. Therefore, trade associations became the primary mechanism for organizing various industries in the United States.

Typically, industry trade associations engage in a variety of activities to improve the performance of the market. Among these activities are market surveys; the promotion of new uses of products or of new products themselves; commercial and industrial research; the collection and distribution of industry-wide statistics; the establishment of commercial and technical standards and practices; and uniform cost accounting procedures. Additionally, they usually publish a trade journal, magazines, or other publications; provide job placement within the industry; engage in arbitration agreements and industrial relations activities; arrange group insurance; share the cost of industry advertising and publicity; lobby various governmental agencies; and, in some instances, report prices and sales. Finally, they generally provide some type of codes of ethics and fair practices standard for their members. In all of this, however, the fundamental question is one of what constitutes cooperation and what constitutes conspiracy. In most instances, the answer varies depending on the current level of enforcement of antitrust laws at the federal level.

The handloading of railroad cars. The logs were moved down the mountainside via the ground slide to the right and were then manually rolled onto the log car. The individual to the far left is holding a grab skipper used to release the grab hooks from the logs. The two woodhicks in the center foreground are holding peaveys, used to roll and to lever the logs. Note the men's clothing, even though this is a winter/early spring season.
Photo courtesy of Pennsylvania Lumber Museum

The heyday of trade associations occurred with the coming of the First World War and the 1920s. The prevailing view was that the country was moving from an age of competition to an age of cooperation, and various trade associations were used as industry-wide coordinators by the War Industries Board as they attempted to mobilize the country for war. In the 1920s, given the very positive effects from the war mobilization, the trade association movement was strengthened even more as the Supreme Court finally appeared to approve association activities as acceptable behavior under the antitrust legislation.

These efforts were furthered by the activities of Herbert Hoover, first as secretary of commerce and later as president, as he attempted to make the trade association movement one of the cornerstones of his efforts to give the federal government a role in guiding the economy. He saw the broader perspective provided by the trade association and the concerted efforts that these associations could bring to the marketplace as a positive benefit to the individual firm as it attempted to plan for the future. To Hoover, the trade association was one mechanism to be used in modifying the business cycle. He saw the role of the trade association as collecting and disseminating statistics and interpreting the trends to member firms as well as promoting standardization and improved management.

The Great Depression ended Hoover's dreams of establishing a managed economy through the use of trade associations. However, trade associations had a reprieve under the New Deal with the passage of the National Recovery Administration (NRA) in 1933. Under the NRA, and building on the experience from the First World War as a way of "mobilizing" business to fight the Depression, trade associations were once again used to bring firms together to write "codes of fair competition" that fixed prices, established minimum wages and maximum hours of work, and, in some instances, allocated markets. The codes were enforced by "code authorities" who were usually the trade associations; the associations were paid a fee for administering the codes. Thus, the federal government was subsidizing the associations. The NRA became the keystone of the New Deal's efforts to fight the Depression until it was declared unconstitutional by the Supreme Court in 1935.

After this time, trade associations had a much harder time functioning in an increasingly hostile federal environment, and lobbying efforts had to be mounted in an effort to provide legislative relief for various industry trade associations. At the same time, however, by the 1940s, trade associations had become an integral part of the American economy. Trade associations today, along with other business groups, probably have, proportionally, "greater influence than any other identifiable group in American society."

RETAIL LUMBER TRADE ASSOCIATIONS

Lumber trade associations have a long pedigree in American History and generally follow the history of trade associations in America. Within the lumber industry itself, there are three types of trade associations: lumber manufacturers, wholesalers, and retailers and their history is intertwined. However, for the sake of brevity, we will deal only with the retail associations.

The retail lumber trade associations, from the standpoint of membership, represent the largest organized element within the lumber industry and generally represent the oldest organized cooperative efforts within the lumber trade.

The earliest record of a retail association is that of the Iowa Retail Lumber Dealers' Association which was founded in 1876 in an effort to stop "wholesalers from shipping carload lots of lumber to a consumer for the same sum charged the retail dealer in the same town." The Iowa

Retailers agreed that if any wholesaler shipped a carload of lumber to anyone outside of a dealer, in any town where there was a member of the association, that member would notify the secretary and the latter would draw upon the wholesaler who shipped the lumber a sight draft for $10. The secretary was to send this amount to the member whose business was being cut. If the draft was not honored the association was pledged to a man to buy no more lumber from the offending wholesaler. This plan worked so well, apparently, that it was copied throughout the Midwest. The Iowa association only had a short history and it and its successor organization were disbanded in 1890.

However, state and regional associations began to develop throughout the country during this time and, generally, these new associations had the word *protective* in the title. H. C. Searce, secretary of the Illinois Lumber and Builders' Supply Dealers' Association, speaking about the purpose of these new associations, said: "The original and almost sole purpose of the retail trade association was to protect its members against the sale of lumber directly to consumers by the wholesalers. Big stick methods were used and justifiably so."

By the turn of the century, protection was needed from a new threat—mail-order houses which threatened to make serious inroads on the retail lumber industry, especially the rural lumber yards. Throughout the early history of the retail lumber associations, mail-order competition would continue to be seen as a threat to the local lumber dealer. Today the equivalent is modular construction.

Along with the various state retail trade associations, the lumber industry also began to create a number of regional and national organizations to speak for the retail lumbermen. One of the earliest was the United Association of Lumbermen, which was organized in 1890 to bring together the officers of the various retail associations. By the late 1890s, this had been merged into another group called the Lumber Secretaries' Association which was comprised of the secretaries of the various associations within the lumber trade. By 1900 a Retail Lumber Secretaries Association was created to coordinate the activities of the various state associations and this period also saw the development of various regional associations, such as the Eastern States Retail Lumber Dealers Association, which provided the crucial Supreme Court case mentioned earlier.

Today, these various state efforts have been centralized in organizations such as the Eastern Building Material Dealers Association which represent members in several states as well as in national organizations such as the National Lumber and Building Materials Dealers Association.

Note the presence of oxen in this early logging scene of a decked landing. Probably a rail line was to be laid, although this was usually done before cutting.
Photo courtesy of Pennsylvania Lumber Museum

A Barnhart log loader, owned by the Goodyear Lumber Company, loading log cars. The log loader was among the first mechanized labor-saving devices used in logging.
Photo courtesy of Pennsylvania Lumber Museum

Typical logging camp scene. Photo courtesy of Pennsylvania Lumber Museum

Typical logging camp group scene. Given number and size of buildings, a large camp. Note the sawfiler's wooden saw clamp on the hillside to the left of the first building. The scaler, with log scale stick, is standing to the far left.
Photo courtesy of Pennsylvania Lumber Museum

*Tram road in Centre County, 1894.
Photo courtesy of Pennsylvania Lumber Museum*

*Building a split log slide. These slides could be iced in winter and greased in summer.
Photo courtesy of Pennsylvania Lumber Museum*

Log drive on Little Kettle Creek, Potter County, circa 1880. Log drives took place before the invention and practical adaptation of the geared logging locomotives. Drives took place during the spring months with high water. This work was extremely injurious to animals and to drivers due to the water temperature.
Photo courtesy of Pennsylvania Lumber Museum

A small log holding pond. The pond man is sorting the logs.
Photo courtesy of Pennsylvania Lumber Museum

*Log drive on the Little Pine Creek.
Photo courtesy of Pennsylvania
Lumber Museum*

*Round timber raft on the West Branch
of the Susquehanna River. This is a
three-platform raft. Note raft sweeps at
fore and at aft. Rafts went down the
Susquehanna River to Fort Hunter,
just north of Harrisburg.
Photo courtesy of Pennsylvania
Lumber Museum*

Scene from a stereoscopic card depicting a rafting-in area for square timber rafts.
Photo courtesy of Pennsylvania Lumber Museum

"Shield Bros' Champion Load. Six green pine logs drawn from skidway to Weston Bros' Pinery on Honeoye Creek, Pa., a distance of over two miles, to banking, by one team, owned and driven by Mr. Jas. Shields. Estimate weight, 20 tons."
Photo courtesy of Pennsylvania Lumber Museum

Sawmill scene, possibly Sweden Valley, Pennsylvania.
Photo courtesy of Pennsylvania Lumber Museum

Drying yard, Keating Summit, Pennsylvania.
Photo courtesy of Pennsylvania Lumber Museum

Sawmill drying yard with stickered boards. Unusually long length lumber in foreground. Probably, I. M. Edgecomb and Sons, Knoxville, Tioga County, Pennsylvania, circa 1913.
Photo courtesy of Pennsylvania Lumber Museum

Lumberyard at Masten, Pennsylvania, circa 1916–1917.
Photo courtesy of Pennsylvania Lumber Museum

Photo courtesy of Pennsylvania Lumber Museum

Photo courtesy of Pennsylvania Lumber Museum

Ruins of the Emporium Lumber Company mill at Keating Summit, Pennsylvania, circa 1901. Mill was established in 1893; burned in 1901; rebuilt and operated until 1913. It was a single band-mill that produced 25-30,000 feet during an eleven-hour shift. Products manufactured included lumber, railroad ties, chair rungs, piano keys.
Photo courtesy of Pennsylvania Lumber Museum

A veneer saw in a sawmill.
Photo courtesy of Pennsylvania Lumber Museum

*"The Loup Up and Down Sawmill" near Loysville, Perry County, Pennsylvania, 1921.
Photo courtesy of Pennsylvania Lumber Museum*

*Local and regional lumber associations were in existence before the establishment of state associations. This is a group picture of the members of the Lumber Exchanges of Philadelphia and Norfolk at a meeting of the two groups in Norfolk, Virginia, 1891.
Photo courtesy of A. K. Shearer Company*

A. K. Shearer Company, founded in 1863, an EBMDA centennial company, has an unusually fine collection of photographs cataloging its history. (Shown on the following pages) This page: An 1867 photo of Moyer and Shearer's Lumber Yard and Hay Press. Photos courtesy of A. K. Shearer Company

An 1888 photo showing A. K. Shearer's Lumber Yard with Mr. Shearer in the doorway of the office.

An 1894 photo showing A. K. Shearer on the left.

A 1900 photo showing Mr. Shearer on the right.

A line drawing of the A. K. Shearer Lumber Yard which was used in advertising around the turn of the century.

South Street and Schuylkill River, Philadelphia, Pennsylvania, circa 1900. Photo courtesy of Pennsylvania Lumber Museum

Chapter II
The Early Years

They met at 10:30 a.m. in the Common Council Chamber of the City Hall in Reading, Pennsylvania, on March 22, 1892. They represented fifty-three firms and established a temporary organization. James A. O'Reilly was elected president, George W. Randenbush secretary, and George F. Lancaster assistant to the secretary. The immediate order of business was the creation of a permanent structure, and by 11:00 a.m. a five-member committee on permanent organization and a ten-member constitutional drafting committee had been created. The members retired to begin their work. "The time was occupied by the others present discussing matters generally until 12 o'clock when the meeting was recessed until 1:30 p.m."

After lunch the meeting was reconvened by President O'Reilly to hear the reports of the two committees. However, before business could be conducted, "it being noticed that several persons were present who should not be, they were invited to retire by the President and did so in good order."

The Committee on Permanent Organization proposed that the organization should be known as the "Pennsylvania Lumbermen's Protective Association"* with the affairs of the group managed by a president, a vice president, a secretary/treasurer, and a board of nine directors.

It should be pointed out that, throughout the early years of the organization, the title Pennsylvania Lumbermen's Protective Association and Pennsylvania Lumberman's Protective Association were used interchangeably. For consistency, Pennsylvania Lumbermen's Protective Association will be used throughout this work.

PENNSYLVANIA LUMBERMEN'S PROTECTIVE ASSOCIATION
FOUNDING MEMBERS: 1892–1894

x1. Steelton Planing Mills, Steelton
x2. D. H. Taylor and Sons, Morrisville *#
x3. Sadler and Kauffman, Harrisburg
x4. Sturdevant and Goff, Wilkes-Barre
x5. B. B. Martin and Co., Lancaster
x6. Harrisburg Planing Mill, Harrisburg
x7. John Bell, Harrisburg
x8. Atticks and Britcher, Dillsburg +
9. George W. Langlets, Harrisburg *#
x10. Reinoehl Lumber Co., Lebanon
x11. John Benore and Sons, Scranton
x12. J. B. Kauffman, Steelton#
x13. Lancaster Planing Mill, Lancaster
x14. Sener Bros. and Co., Lancaster
x15. Merritt Bros. and Co., Reading
x16. Harrison Ball, Mahanoy City
x17. G. Sener and Sons, Lancaster +
x18. Mason and Snowden, Scranton
x19. Jacob Beitzel, York
x20. J. H. Sterner, Schuylkill Haven *#
x21. Riley Bressler, Tower City #
x22. Ball and Millington, Tremont *#
x23. George Fleisher, Newport #
x24. George Ball and Co., Minersville *#
x25. Conrad Lee, Wilkes-Barre
x26. A. Ryman and Sons, Wilkes-Barre
x27. Sol S. Lenhart and Sons, Harrisburg #
x28. Spruks Bros., Scranton
x29. Haney White and Co., Philadelphia
x30. I. F. March and Sons, Bridgeport #
x31. Fox and Graeff Co., Lebanon
x32. Reinoehl and Nutting, Lebanon
x33. Olyphant Lumber Co., Olyphant +
x34. Ezra Finn and Sons, Scranton
x35. P. C. Horrine, Reading #
x36. Gasser and Johnson, Reading
x37. Kline and Shunk, Reading
x38. George F. Lance, Reading
x39. F. P. Heller, Reading
x40. J. Mengle Fisher, Pottstown
x41. J. Frank Althouse, Pottstown *
x42. P. L. Egolf and Son, Pottstown *#
x43. R. A. and J. J. Williams, Philadelphia
x44. E. W. Trexler and Son, Allentown
x45. West Reading Planing Mill, Reading *
x46. Price and Howarth, Scranton
x47. Miller Cilley and Co., Lebanon
48. C. F. Brown, South Bethlehem
x49. Steeley Weinhold and Co., Reading *#
50. [John F.] Taylor and Co., Taylor #
51. W. B. Hull, Scranton
52. Joseph Ansley and Son, Scranton
53. Peck Mfg. Co., Scranton
54. Plymouth Planing Mill Co., Plymouth #
55. Zearfoss and Hilliard, Easton

56. H. W. Ahlers and Co., Allegheny City
57. Kendig and Lauman, Middletown
58. Pottsville Lumber Co., Pottsville
59. Charles Young, Hanover
60. Hugh Burke, Dunmore *#
61. J. L. Chapman, Scranton #
62. Abele Bros., Scranton *#
63. A. J. Bassler, New Ringgold #
64. J. E. Patterson, Wilkes-Barre
65. M. D. Brown and Co., Olyphant #
66. Morgan Planing Mill Co., Wilkes-Barre
67. George S. Beeten and Co., Carlisle #
68. Wyoming Valley Lumber Co., Pittston
69. J. T. Pethick and Bros., Carbondale
70. Green Ridge Lumber Co., Scranton
71. Thomson and Son, Luzerne
72. Thomas Kerns, Slatington
73. Mills, Baker and Co., Carbondale
74. T. C. Robinson, Carbondale *#
75. Guest, Grater and Co., Norristown
76. J. F. Hazard and Co., Philadelphia
77. J. W. Craig, Chambersburg
78. J. S. Allam, South Bethlehem
Associate Member - Henry Stephens, St. Helen, Michigan

New Members as of 1893:
79. J. J. Hess, Hellertown
80. Keck and Bros., East Allentown
81. Reuben Transue, South Bethlehem
82. Wm. Wohlsen, Lancaster
83. H. Lanius and Sons, York
84. Borhek and Miksch, Bethlehem
85. Joseph Milleisen, Mechanicsburg
86. Billmeyer and Small, York
87. Fitzgerald, Speer and Co., Pen Argyl
88. Paul Barrall, West Nanticoke
89. Dershimer and Griffin, Pittston
90. Mulherin and Judge, Scranton #
91. D. W. Hess, Scotland
92. East Stroudsburg Lumber, East Stroudsburg

x At organizing meeting, March 22, 1892
* No membership for 1893
# No membership for 1894
+ Notation in margin "went out of business"

At organizing meeting, March 22, 1892, but did not join organization:
W. H. Schall, Barto
W. F. Dreibelbis, Royersford
A. K. Deysher, Reading
J. D. Kocher, Orwigsburg
Samuel Seaber, Lititz
R. M. Taylor, West Chester

The Constitution Committee reported next and proposed a constitution that was agreed to by the assembled group. In the words of the Constitution, the object of the Pennsylvania Lumbermen's Protective Association was "the protection of its members against sales by wholesale dealers and manufacturers to consumers and the giving of such other protection as may be within the limits of co-operative association." The organization was to be headquartered at Reading, although this was changed a year later to establish the headquarters wherever the secretary resided.

Membership was defined in Article II of the Bylaws of the Constitution as "any person who may be regularly in the retail Lumber Trade and owning or operating a Lumber Yard or Planing Mill engaged in retail trade in this state." The Constitution defined retail lumber dealer as "an individual firm or corporation occupying a yard to receive, sort and store and from which to sell and deliver Lumber with the provision that any difference existing between planing mill men and retail Lumber dealers shall be settled between local dealers and mill men." They also provided for associate memberships for out-of-state individuals owning or operating a retail lumber yard.

Additionally, this initial constitution provided for the management of the association by a president, vice president, secretary/treasurer and a nine-member Board of Directors, all of whom were to be elected at the annual meeting to be held in January. The officers of the association were ex-officio members of the board, and the board was given authority to appoint committees, disburse funds and "print and circulate documents in the interest of the Association and devise and carry into execution such other measures as they may deem proper to promote the objects of this Association," and the Bylaws specifically charged the secretary with preparing, printing, and distributing a membership list. There were to be two regular meetings a year to be held the second Wednesday of January and July each year, and the annual dues were to be $5.00.

Probably the most important provision of the new association was Article IV of the Bylaws of the Constitution. It provided that

> whenever and as often as any manufacturer or wholesale dealer in timber, lumber or mill work shall directly or through commission merchants, agents, brokers or scalpers quote prices, furnish price lists, solicit sales, sell lumber, timber or millwork to any person not a regular dealer, it shall be the duty of any member of this Association having knowledge of such transaction to notify the Secretary in writing, stating all the particulars possible and the Secretary upon the receipt of said notice from

Page from minute book showing minutes from the morning session of the organizational meeting. The page shows the establishment of a Committee on Permanent Organization and a Committee on Constitution and Bylaws.
Photo courtesy of EBMDA

any member shall notify each member of the association and all members then agree to refuse to buy from said wholesale dealer until notified by the Secretary that they have pledged themselves to cease selling to customers. And upon the refusal of any member of the Association to heed the notice of the Secretary to cease buying from said wholesale dealer, he shall be expelled from the Association.

This provision was the very heart of the association, the reason for its creation, and would be the main business in the early years of its existence, but would also lead to legal problems in the future.

Finally, the organizational meeting agreed to cooperate with any similar association in the United States, and they elected the initial slate of officers: F. P. Heller of Reading as president; S. H. Sturdevant of Wilkes-Barre as vice president; George F. Lance of Reading as secretary/treasurer; and to the Board of Directors, J. L. Kaufman, Harrisburg; H. C. Trexler, Allentown; T. J. Snowden, Scranton; C. Haney, Philadelphia; H. K. Baumgardner, Lancaster; and Jacob Beitzel, York.

Thus began the Pennsylvania Lumbermen's Protective Association (PLPA). During 1892, 78 individuals and firms paid the $5.00 dues and became members. In 1893, 92 names appeared on the membership list, although 13 members from the previous year did not pay their dues, with 2 of those 13 having gone out of business. By 1894, 144 members had joined, but, again, non-payment of dues continued to be problem, as it would throughout the early years of the association when 24 of the original members failed to pay their dues and 6 of those went out of business, died, or withdrew their membership. Thus, the association began with 78 members in 1892, went to 79 in 1893 and 119 in 1894. An analysis of the geographical location of the early members shows that, while a few came from Philadelphia, the strength of the organization was in an arc around the city that went from the Maryland border going through Lancaster, Reading, and Harrisburg and up into the coal region including such cities as Allentown, Wilkes-Barre, Scranton, and Hazleton.

The minutes of the early meetings of the Board of Directors make it very clear that one of the main reasons for the establishment of the Pennsylvania Lumbermen's Protective Association was to protect the markets of retail lumbermen from incursion by fly-by-night operators or scalpers and wholesalers selling directly to consumers. Forty-three shippers were reported as having violated Article IV of the Bylaws in 1892. Of that number, four were excused, six provided satisfactory explanation of the circumstances under which they had made the questionable transaction, and the minutes reveal that thirty-three were "reported." However, once the association was able to establish some legitimacy to its operation, the number of shippers reported for violations of Article IV began to diminish. In 1894 there were twenty-seven; in 1895 there were fifteen; and in 1896 there were sixteen. While this would be a continuing issue throughout the early years of the association, and one that the Board of Directors would repeatedly discuss, the list became smaller and smaller, and usually cases that were reported to the board would be handled by the Executive Committee. These were usually explained or excused. Also, the Executive Committee would consistently communicate with surrounding state associations to obtain their views on many of the cases since they involved companies from out of state or firms engaged in interstate commerce. Throughout the 1890s, approximately twenty-five firms would be "on report." To illustrate that things do really change that much, the minutes of the Executive Committee for January 27, 1893, report that the committee received a communication from S. J. Straus, Esq., "providing his legal opinion that the Association had legal liability for placing names of parties on our list for violation of our by-laws."

It is interesting to note that in the first year of operation, the total expenses of the treasurer were $392.09. Of that amount, the largest single expense was for the equipping of the association's office. The cost of desk and chairs was $77.25. The second largest expense was the printing of the Constitution and Bylaws, which came to $48.85. However, the association thought on a grand scale, since they paid to have 1,000 copies published. During the first year, the secretary was paid $100; the total postage charge was $12.50; and Board of Directors' meetings cost $7.00 each. By the second year, the secretary was paid an annual salary of $150.00 and postage had jumped to $37.02 with total expenses coming to $390.63.

Listing from minute book showing some of the original members of the Pennsylvania Lumbermen's Protective Association.
Photo courtesy of EBMDA

CONSTITUTION, BY-LAWS

AND

LIST OF MEMBERS

OF

THE PENNSYLVANIA LUMBERMEN'S PROTECTIVE ASSOCIATION.

1895-'96.

FRANKLIN PUBLISHING CO.
4 PEARL ST., BOSTON.

Title page from the first publication of the association, printed in 1895. Photo courtesy of EBMDA

Almost immediately upon the establishment of the association, the members began to address a variety of issues affecting its development and how the efforts of the Pennsylvania Lumbermen's Protective Association (PLPA) might assist in the operation of their businesses. The first several years of the association's history saw a number of initiatives that would establish a pattern and set a course for its entire history.

At the first annual meeting in January 1893, the association created several committees to more effectively deal with issues. A Committee on Transportation was created and immediately began to work to memorialize railroads on the method of payment for the rental of railroad cars. A Committee on Legislation was created to monitor the activities of the Pennsylvania legislature and it very quickly began to lobby for a change in the mechanics lien laws in Pennsylvania. At the semi-annual meeting held in July 1893, a Committee on Increasing Membership was established. This was done after it was reported that the secretary had to send dunning notices to delinquent members. At the same meeting it was agreed to solicit funds from members to hire a detective for two months to investigate firms using fictitious names and secret manifests in the Williamsport area. It was felt that, if they could effectively stop such an activity, membership would increase dramatically.

At the same time, the association took steps to become involved with other state lumbermen's organizations. The common practice was to elect the officers of the surrounding state associations as honorary members of the PLPA. In this way, close contact was quickly established with the New Jersey, New York, and Connecticut associations. Almost immediately after the creation of the PLPA, the group began to develop contacts with these state associations and the national lumbermen's association by sending members of the Executive Committee to state meetings. In July 1893, the PLPA named the *New York Lumber Trade Journal* as the official organ of the association. For a fee of $1.00 per member PLPA was allowed to send material for inclusion in the *Journal*.

An issue that divided the association early in its history was the inclusion of wholesale dealers and manufacturers as members. This apparently was an issue for all of the lumbermen's associations, and a meeting of representatives of the state associations of Pennsylvania, New York, Massachusetts, Connecticut, and the National Lumbermen's Association was held in Newark, New Jersey, on November 21, 1893. It was agreed to meet with the Wholesalers Committee on December 19, 1893. At that meeting, common definitions of wholesaler and retailer were agreed upon and each of the associations proceeded to include these definitions in their Bylaws. A retailer was defined as "those persons within those states who purchase Lumber from Manufacturers and recognized wholesale dealers, who own or operate a lumber Yard where a stock of Lumber is at all times kept for sale to the public and who are not themselves engaged in building or otherwise consuming Lumber in competition with the usual customers of a retail lumber yard." Where questions arose, an arbitration procedure was established with the Wholesalers Association, and each state association agreed to add a provision to their Bylaws that called upon their members to purchase only from those who subscribed to these common definitions. Each of these provisions was added to the PLPA Bylaws at the 1894 annual meeting. At the same meeting, it was agreed to employ a "suitable person to solicit additional memberships" and it was agreed to appoint a committee to meet with the Philadelphia Lumberman's Exchange in an effort to have them join the PLPA.

By the July 1894 Semi-annual meeting, it was reported that the Philadelphia Lumbermen's Exchange was in favor of merging the two organizations with the headquarters to be located in Philadelphia. The officers of PLPA were instructed to work toward such a merger. At the same meeting, PLPA agreed to join the United Lumberman's Association, the national lumberman's association, and to join with the other lumberman's associations in the state (Allegheny County and Philadelphia Lumberman's Exchange) to attempt to get an effective mechanics lien law enacted. Finally, the Association created a committee to report on the feasibility of creating an insurance company for lumber dealers within the association because of the high premiums paid by dealers in the state. In conjunction with the Philadelphia Lumberman's Exchange, this preliminary study would lead, a year later, to the creation of the Pennsylvania Lumbermen's Mutual Fire Insurance Company, for "lumber yards, woodworking establishments and members of different lumber associations." Within a year of its creation, the company reported having written over $1,000,000 in insurance, $14,000 paid premiums, and no losses. The report from the Pennsylvania Lumbermen's Mutual Fire Insurance Company was a continuing success story at the early annual meetings and would eventually be pointed to as an example of the value of united action.

The creation of the fire insurance company was not the only early success. The association's efforts, along with the efforts of other associations on behalf of a curative mechanics lien law, were also successful. In particular, working in conjunction once again with the Philadelphia Lumberman's Exchange, who was to name a senator to take charge of the bill in the state Senate and "we will name a Representative to handle the bill in the House," they successfully passed a mechanics lien law

CONSTITUTION.

ARTICLE I.

The title of this organization shall be "THE PENNSYLVANIA LUMBERMEN'S PROTECTIVE ASSOCIATION," and it shall have for its object the protection of its members against sales by wholesale dealers and manufacturers to consumers, and the giving of such other protection as may be within the limits of co-operative association; the town or city in which the Secretary and Treasurer resides, shall be considered the headquarters of the Association.

Page from the 1895–96 "Constitution and By-Laws."
Photo courtesy of EBMDA

BOARD OF DIRECTORS.

S. H. KECK,	Allentown
J. W. CRAIG,	Chambersburg
GEO. F. LANCE,	Reading
W. Z. SENER,	Lancaster
W. M. JAMES,	Steelton
E. M. WILLARD,	Philadelphia

Listing of committees and members from 1895–96 "Constitution and By-Laws." (above and on following pages)
Photos courtesy of EBMDA

EXECUTIVE COMMITTEE.

S. H. STURDEVANT, *President*,
. . . . Wilkes-Barre, Pa.
O. M. BRANDOW, *Vice-President*,
. . . . Wilkes-Barre, Pa.
T. J. SNOWDON, *Secretary and Treasurer*,
. . . . Scranton, Pa.

COMMITTEE ON LEGISLATION.

S. H. STURDEVANT, *President*,
. . . . Wilkes-Barre, Pa.
O. M. BRANDOW, *Vice-President*,
. . . . Wilkes-Barre, Pa.
T. J. SNOWDON, *Secretary and Treasurer*,
. . . . Scranton, Pa.

COMMITTEE ON ENLARGEMENT OF ORGANIZATION.

LESLIE S. RYMAN, . Wilkes-Barre, Pa.
J. E. PATTERSON, . Wilkes-Barre, Pa.
E. M. WILLARD, . Philadelphia, Pa.

---

more to their liking. By April 1895, the association was expressing its thanks to Senator Penrose and Representative Farr, as well as to the governor, for their assistance in passing the lien law. They had learned an early lesson in the value of lobbying the legislature and in having members contact their respective legislators to express their views.

United action was also seen as being beneficial for the association movement in general, and PLPA was an early supporter of regional cooperation among lumber associations in the eastern part of the United States. In conjunction with the other lumbermen's associations, meetings were held to discuss regional issues. From these meetings came arrangements to have arbitration agreements, common lists of malefactors, central printing of publications to save printing costs, periodic meetings of the secretaries of the associations, and common positions regarding issues before the United Association of Lumbermen.

By the end of 1896, the secretary/treasurer reported a membership of 165, total expenses of $1,141.62, and the first charge for a telephone. However, the membership figures were always a source of debate at any meeting of the Board of Directors or at the annual meeting. As the turn of the century approached, the association found itself losing members rather than growing. In 1896 Charles B. Keller was hired to solicit new members at the rate of $50.00 per week for the time he was actually employed in soliciting members. By 1897, the membership figures had dropped to 101 members and the Executive Committee passed a resolution making all "legitimate retail lumber dealers of the State of Pennsylvania not members of the association be elected to membership for the year 1897 without the payment of dues" (which remained at $5.00). This immediately increased the membership to 150. However, by the 1898 Annual Meeting, it was reported that only 98 dues-paying members belonged to the association.

One of the mysteries of the early years of the association is the method of communicating with members. With the exception of the Constitution and Bylaws, which were printed as a book with advertisements and the pictures of all the officers and directors, none of the early publications exists. It is clear from the records of the association that some type of publication was being sent to members. While reference has already been made to arrangements for publication of notices in the *New York Journal*, it is unclear what other communications were being used. Whether there was some type of formal newsletter, a circular letter, or some other type of publication remains a mystery. However, we do know that they published some type of "report," if for no other reason than to list those companies who had been censured by the association and also to publicize those retail lumbermen which were members of the association. In addition, after discussion with several different organizations, including the various state lumbermen's organizations, both retail and wholesale, the association decided to publish a list of retailers in the state for the assistance of wholesalers. Finally, to further document this communications network, the secretary of the association wrote the following to members who had not paid their dues in 1898: "For the past year and a half the Secretary. . . . has mailed to your address our regular reports relating to the abuses practiced by said wholesale shippers hoping that the reports would be appreciated by you to the extent of your taking an interest in the workings of the Association and becoming a member thereby aiding both yourself and the retail trade generally in eradicating the evils which the trade is subject to." Unfortunately, except for the 1895-96 Constitution and Bylaws, none of these publications exist.

One of the more vexing problems for the Association, as well as for other lumbermen's organizations, was establishing a working relationship with wholesale lumbermen and, in particular, coming to a common definition of wholesaler and retailer. Efforts to deal with the problem were undertaken by PLPA itself, as well as the other lumbermen's associations, including an organization that PLPA would eventually join in 1903, the Eastern States Retail Lumber Dealers Association. Another group that took the lead in attempting to organize a meeting of all the groups involved was a newly organized "Secretary's Association" which was composed of the secretaries of the various state and regional associations, and which the PLPA helped to organize. This meeting between the regional lumbermen's associations and the National Wholesale Lumber Dealers Association took place in Boston, March 1–2, 1899, and the "Boston Agreement" was concluded. Common definitions were agreed to and the trade was classified into three sections: manufacturers, wholesale dealers and agents, and retail dealers. After some continuing disagreements on interpretation of the "Boston Agreement," an arbitration procedure with the National Wholesalers Association was arranged and the PLPA Bylaws were amended in January 1900, to conform to the new arrangement. Once these were implemented the issue of wholesaler versus retailer very rapidly declined as an issue dealt with at the annual meetings, the Executive Committee or Board of Directors meetings. Instead the issue became one of how to classify businesses into one of the four categories recognized by the association: Class A, Retail Lumber Dealers and "trade for the Wholesalers to sell to;" Class B, Wholesale Consumers, Manufacturers and Trade for the Wholesaler; Class C, Carpenter, Contractors, "who are trade for the Retailers only;" and Class D,

---

COMMITTEE ON INSURANCE.

Geo. F. Lance, . . Reading, Pa.
S. H. Keck, . . . Allentown, Pa.
W. Z. Sener, . . . Lancaster, Pa.

COMMITTEE ON COMPLAINTS.

S. H. Sturdevant, *President*,
. . . . Wilkes-Barre, Pa.
O. M. Brandow, *Vice-President*,
. . . . Wilkes-Barre, Pa.
T. J. Snowdon, *Secretary and Treasurer*,
. . . . Scranton, Pa.

COMMITTEE ON RAILROADS AND TRANSPORTATION.

J. W. Howarth, . . Scranton, Pa.
J. E. Patterson, . Wilkes-Barre, Pa.
W. Z. Sener, . . . Lancaster, Pa.
G. F. Lee, . . Wilkes-Barre, Pa.

Pictures of officers, directors, ex-directors of PLPA from 1895–96 "Constitution and By-Laws." (below and on following pages) Photos courtesy of EBMDA

*S. H. Sturdevant
President*

*O. M. Brandow
Vice President*

*T. J. Snowdon
Secretary and Treasurer*

"Consumers, and trade for the Retailers only."

Concurrent with this issue of classification, the association dealt with several other issues of concern to the retail lumbermen, not the least of which were declining membership, railroad practices, forestry conservation, and legal issues including incorporation.

1900–1910

As indicated by the minutes of the annual meetings and Board of Directors meetings, fluctuation in membership and the need to continually work at increasing the size of the association was a constant throughout its history, but especially so in the early years. While there had been an auspicious start, by 1900, membership had dropped to 79 and renewed efforts were undertaken to "enlarge" the membership with members being given blank membership certificates to distribute to prospective members. Through personal contact by members with non-members the values and benefits of the PLPA were explained. These efforts began to pay off immediately, and by 1901, membership increased to 93; to 104 by 1902; to 129 by 1903; and to 149 by 1904. But efforts to enlarge the membership would continue and be a constant topic at annual meetings and meetings of the Board of Directors.

Another reason for the upswing in the membership was the fact that the association hired a paid secretary rather than having a member as secretary. At the January 9, 1902 Annual Meeting, the Bylaws were amended to have the Board of Directors select the secretary, and that afternoon they chose B. F. Laudig at a salary of $200.00 per year. It was agreed at the July 1902 meeting of the Executive Committee that the secretary was to travel to various areas of the state to obtain new members and that he would receive $5.00 for each new member as a way of defraying his travel costs. Laudig was the first paid staff person of the association, and his efforts were immediately successful as shown by the increase in membership. Within six months, he had created a "Bureau of Information" for members to use to obtain information on issues confronting the industry, and he was able to report that he had received over five hundred communications and had sent out a like number.

Railroad practices were also a continuing item of discussion in the early years of the association. Among the issues of concern were railroad rates and, in particular, the practice of shipping lumber and having it milled along the way. Also of concern were forest conservation and the prevention of forest fires. The semi-annual meeting, which was held in the summer, usually met close to state forest areas and attendees were given tours of the forest. The association repeatedly passed resolutions dealing with the lumbering industry and the need to replenish the forests of the state. Additionally, they supported efforts to combat forest fires by calling for the enforcement of existing legislation to prevent forest fires and proposing that railroads establish fire lines along their right of ways as one method to reduce the number of forest fires.

The legal issues are more difficult to explain because the minutes of the association contain only cryptic references to legal action. It would appear that sometime around 1901–1902, the association was sued in county court in Wilkes-Barre, Pennsylvania, apparently by someone who had been "reported." The records of the association were subpoenaed for evidence. As a result, at the July 9, 1902 meeting of the Board of Directors, the president was directed to obtain legal advice regarding the legality of the association's "report" listing. Additionally, in 1903, the process of becoming incorporated was begun as one way of protecting the officers and members from legal action, and the court granted the corporate charter on July 6, 1903. At the semi-annual meeting on July 9, 1903, the corporate charter was ratified by the membership, the PLPA was disbanded and the new corporation, the Pennsylvania Lumberman's Association, Inc., took its place. A new Constitution and Bylaws were adopted and a Board of Directors elected.

However, within a month, F. C. Hanyen, a lawyer, pointed out a number of problems with the new corporate charter as to procedure as well as substance: it was not subscribed to by the proper number of Pennsylvania citizens; it was not advertised properly; the filed copy was filled with corrections; but, most importantly, the stated purpose of the organization was deemed inappropriate. Mr. Hanyen told the Board of Directors, "The most serious defect, as I view the matter, is the purpose for which the charter was granted as set forth in the said instrument. It is stated therein that the purpose of the corporation is for mutual benefits and protection to be supported by funds collected from its members. I do not understand the object of your association to be to form a beneficial society for the purpose of paying weekly and funeral benefits to its members in case of sickness and death yet that is what your charter provides for. From your statement to me, I gather that your purpose is, generally speaking, for the encouragement and protection of trade and commerce. This being true, the charter you have fails to meet your purpose."

While Mr. Hanyen indicated that the granting of the charter "corrected" these defects, but cautioned that if there was ever a show cause order to revoke the charter, these problems would be seen as glaring errors and would cause the corporation serious problems. As a result, the Board of

J.W. Craig
Director

George F. Lance
Director

S. Henry Keck
Director

W. T. Sener
Director

W. M. James
Director

J. F. Hazard
Ex-Direcor

Directors directed that a new set of incorporation papers be prepared and submitted. In the process they learned an early lesson that they would find useful in the future: the need for good legal advice.

The new, corrected corporate charter was submitted and approved by the court of Lackawanna County on January 4, 1904, and the Constitution and Bylaws as approved at the earlier semi-annual meeting were ratified at the 1904 annual meeting and implemented.

Under this charter, the purpose of the Pennsylvania Lumberman's Association was:

> To foster, protect and promote the welfare of persons engaged in the retail lumber business; and for the protection and encouragement of such trade and commerce by combining the intelligence and influence of members against imposition and fraud, as experience may from time to time prove needful, by bringing about great uniformity or certainty in business connections, and by establishing closer ties of business association among the members.

Because this Constitution and Bylaws set the pattern for the organization throughout most of its history, a detailed description is called for. It began with a general statement of purpose:

> We realize the convenience, if not necessity, of the Retail Lumber Dealers to every community and we are interested in the promotion and general welfare and perpetuation of the Retail Lumber Business.
>
> We recognize the right of the manufacturer and wholesale dealer in lumber products to sell lumber in whatever market to whatever purchaser, and at whatever price they may see fit.
>
> We also recognize the disastrous consequences which result to the legitimate Retail Lumber Dealer from direct competition with wholesalers and manufacturers, and appreciate the importance to the Retail Dealer of accurate information as to the nature and extent of such competition where any exists, and recognizing and appreciating the advantages of cooperation in securing and disseminating any and all proper information for our mutual convenience benefit or protection.
>
> ...The title of this organization shall be "The Pennsylvania Lumberman's Association, Inc." and shall have for its object to secure and disseminate to its members any and all legal and proper information which may be of

interest or value to any member or members thereof in his or their business as Retail Lumber Dealers and the giving of such protection as may be within the limits of cooperative associations. Any person, firm or corporation regularly engaged in the retail lumber trade and owning or operating a lumber yard or planing mill in this state reasonably commensurate with the demands of his community shall be considered eligible to membership in this Association.

The association was to have two regular meetings a year, the second Thursday of January and July, and was to be governed by a nine-member Board of Directors who were to elect a president, vice-president, and treasurer for the association from among their members. Additionally, they were to elect a secretary who did not have to be a member. While the Constitution and Bylaws did not at this time call for regional representation on the Board of Directors (as it would subsequently), it is apparent from the minutes of the board that efforts were made to see to it that board members were distributed throughout the region covered by the association. If for no other reason, this made membership solicitations by board members more convenient.

The Bylaws defined a Retail Lumber Dealer as: "an individual, firm, or corporation occupying a yard to receive, sort, and store and from which to sell, and deliver lumber." Set forth as one of the main purposes of the organization and indicated as one of the duties of membership was the following:

> Any member of this Association having knowledge of a sale by manufacturer or wholesale dealer, commission merchant, agent, broker or scalper, or of quoting prices or furnishing price lists to a customer or to any person not a regular dealer within the territory of such members, may notify the Secretary of the Association in writing, giving as full information thereof as is possible—such as date of shipment and arrival; car number and initial; name used, and such other particulars as may be obtainable—upon receipt of such written notice the Secretary shall immediately verify such report as far as practical and under the direction of the Executive Committee of the Board of Directors shall notify the members of the Association of such sale or sales or shipment by such manufacturer or wholesale dealers, commission merchant, agent, broker, or scalper.

In addition to addressing the incorporation issues and the attendant

H. K. Baumgardner
Ex-Director

J. L. Kaufman
Ex-Director

Harry C. Trexler
Ex-Director

HONORARY MEMBERS.	
Tallapoosa Lumber Co.,	Sistrunk, Ala.
Seaman & Smyth,	Williamsport, Pa.
Guy W. Maynards Sons,	"
Strieby, Sprague & Co.,	"
John Coleman,	"
A. D. Knapp & Co.,	"
Rice & Lockwood Lumber Company,	Springfield, Mass.
Geo. F. Sloan & Bro.,	Baltimore, Md.
H. Snowden & Co.,	Philadelphia, Pa.
Chapman & Hull,	Scranton, Pa.
Haupt Lumber Co.,	"
D. M. Nesbit & Co.,	Lewisburg, Pa.
Geo. W. Robinson & Co.,	Detroit, Mich.
Gulf Red Cypress Lumber Co., (E. S. Davis, Sec'y,)	Trenton, N. J.
Trexler & Turrell Lumber Company,	Ricketts, Pa.
Isaac Frazer,	Harrisburg, Pa.
Emery Lumber Co.,	Williamsport, Pa.
J. C. & W. B. Lance,	Reading, Pa.
Edward Hines Lumber Co.,	Chicago, Ill.

List of honorary members of PLPA from 1895–96 "Constitution and By-Laws." Photos courtesy of EBMDA

organizational details, several other issues were also dealt with. First, a resolution proposed by the National Association of Manufacturers opposing the use of the metric system was adopted. Second, national legislation to give to the Interstate Commerce Commission the power to enforce orders was supported and the Committee on Legislation was instructed to "take what steps in their judgment are necessary to accomplish this result." Third, the association continued to lodge complaints with the publishers of the Lumberman's Credit Association's "Red Book" because of the way they listed contractors as lumber dealers. More care was demanded in their listings. The association would create a committee to investigate this problem and propose a proper course of action.

Throughout the rest of the first decade of the twentieth century the association continued to deal with similar issues. Fortunately, declining membership was not a problem, and by 1910 they boasted a membership of 280. However, attendance at the annual and semi-annual meetings was a concern and the Executive Committee proposed increasing the dues to $15.00 a year with a $10.00 rebate if the member attended the annual meeting. That proposal was defeated by the membership. Other efforts to increase attendance included the creation of a Committee on Entertainment and a motion passed at the 1908 annual meeting urging "that a more interesting program be prepared for future meetings."

On other matters related to the association itself, in 1905 they adopted red and white as their official colors, purchased a typewriter and began to type all of their correspondence and minutes and, in the same year, installed a telephone in their offices and a six-month phone bill of $24 was incurred. In 1910, J. Frederick Martin was hired as secretary at an annual salary of $300.00 plus $2.50 for each new member he obtained. Martin would serve as the chief paid staff officer of the association for more than forty years.

Legal issues continued to concern the association, so they agreed to hire a lawyer on retainer as well as to provide for an observer at a court case in Berwick, Pennsylvania. From cryptic references in the minutes, it would appear that the law was changing and that one of the issues related to how trade associations, such as the PLA, "reported" infractions to their code of conduct. In 1906, the secretary was instructed to follow the legal opinion of Frederick C. Hanyen when issuing the report. In 1908, the secretary was ordered to suspend "publishing of the report of irregular shippers" and the Executive Committee was instructed to revise the Constituion and Bylaws to "bring them into compliance with state and interstate laws."

This was done at the annual meeting in 1909 and, while the Constitution and Bylaws were virtually the same as before, the following Article, entitled "Limitation and Restriction" was included:

> No rules, regulations or by-laws shall be adopted in any manner stifling competition, limiting production, restraining trade, regulating prices, or pooling profits.
>
> No coercive measures of any kind shall be practised or adopted toward any one, either to induce him to join this Association or to buy or refrain from buying from any particular manufacturer or wholesaler. Nor shall any discriminatory practices on the part of this association be used or allowed against any retailer for the reason that he may not be a member of the Association or to induce or persuade him to become such member.

The association's practices, however, must have passed legal muster, because by July 1909, the secretary was directed to "issue an official report monthly to the members, containing all the names of the old firms ordered reported." Additionally, during this time period, the association attempted to stop the practice of "scalping"; or the selling by one retailer to customers in another region covered by association membership.

Increasing the power of the Interstate Commerce Commission (ICC) continued to attract the attention of the membership, and the members were repeatedly called upon to write their congressman and senators on a number of issues related to strengthening the power of the commission. A number of railroad practices particularly irritated the lumbermen, such as charges for car stakes and equipment, the time allowed to unload cars, and, more generally, ticket fees. These were all issues that the association felt a more powerful ICC could address.

National standards were also an issue and the association helped address that issue by calling for national rules for grading, measuring, and counting all woods and timber and for the establishment of a national laboratory for testing. Also, there was increasing concern about the timber supply, and the association repeatedly called upon the U.S. Department of Agriculture to make an estimate of the standing timber of the United States. They also continued their conservation efforts and called upon members to join the Pennsylvania State Forestry Association as a

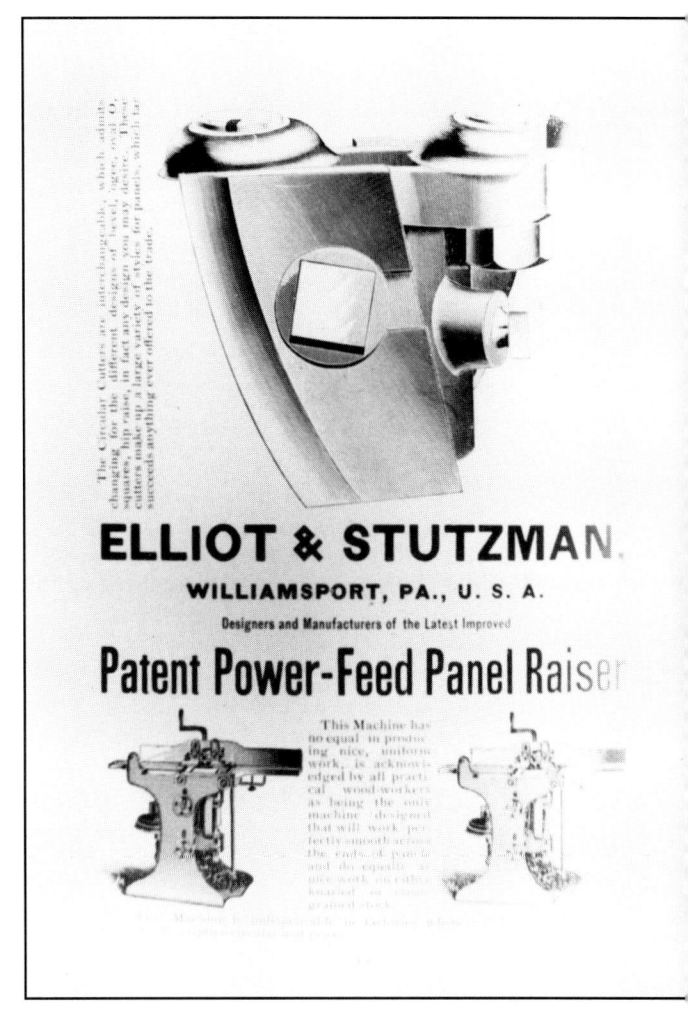

Equipment advertisement from 1895–96 "Constitution and By-Laws." Photo courtesy of EBMDA

way of influencing policy. In 1906, the association went on record in favor of the free admission of Canadian timber to the United States as a way of helping the timber supply.

1911–1920

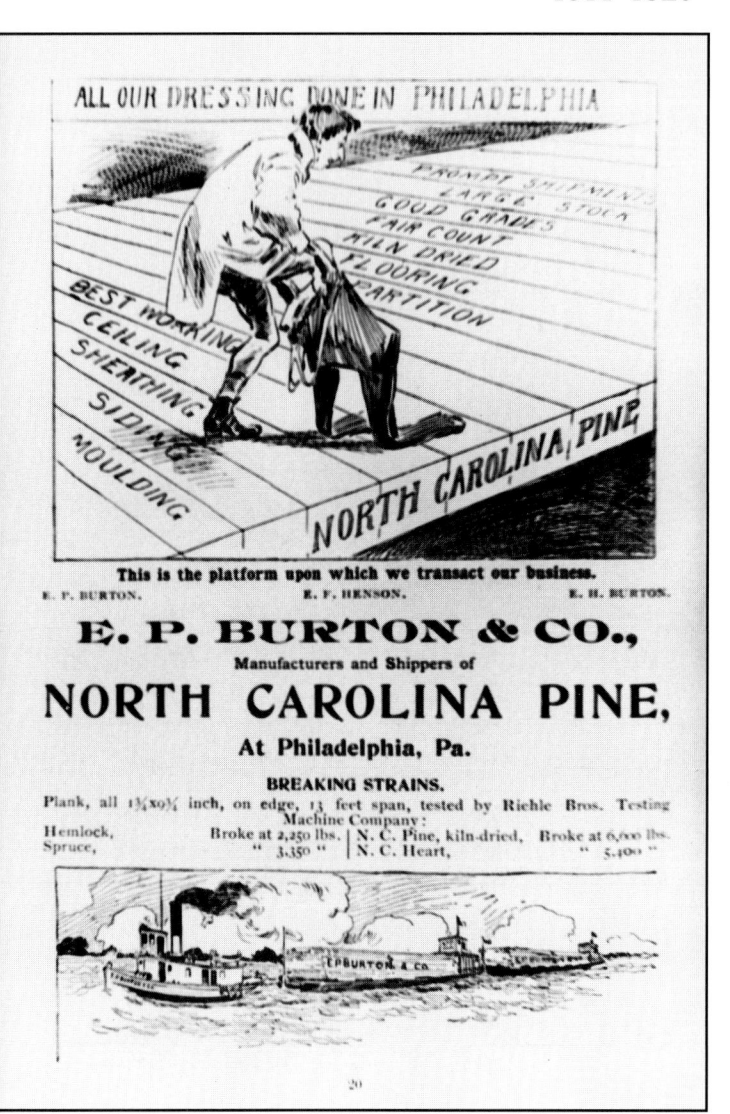

*Advertisement from 1895–96 "Constitution and By-Laws."
Photo courtesy of EBMDA*

The development of a retail lumbermens organization in Western Pennsylvania earlier in the century had been watched closely by PLA, although little interaction occurred. However, in 1910 the association began efforts to bring the western association into closer contact with PLA and, possibly, form a merger. At a special meeting of the Board of Directors, which apparently was called for this purpose, the secretary was instructed to "write the Retail Lumber Dealers' Association of Pittsburgh, requesting the President of that Association to appoint a committee of three of its members to meet a like committee of our Association, at any place they may designate, to consider matters of common interest." From this contact came a joint meeting of the two groups at Bedford Springs in July 1911. The records are skimpy about the actual business transacted. However, it would appear that the meeting consisted of presentations by PLA on the benefits of membership in the association. Nothing came of the meeting and, while there would be continued contact, an actual merger would not occur until 1972. The impression one gets from reading the records is that travel and communication constraints were the major reasons for inaction.

While the minutes are scarce on the actual discussions held at Bedford Springs, they are rather full on the athletic activities taking place at the meeting:

After the meeting baseball nines representing each association retired to the ball field and engaged in a Titanic struggle for supremacy in this particular sport. The game resulted in a one-sided contest in favor of the Western Association, the score being 14 to 2 in their favor. The chief feature of the game was the fielding of C. Frank Williamson of Media who composed the entire outfield and the only misfortune which marred the brilliancy of his work was the fact that he was most likely to be in deep left when a ball was batted to the right field. Fred S. Pyfer worked behind the bat for our team and covered himself in much glory and dropped balls, chiefly the latter.

Declining membership was probably the most important issue facing the association in the second decade of the twentieth century. In 1910 it was reported that 274 members were on the rolls, although 32 were in arrears for dues payments. One year later that number had dropped to 192, and by 1915 to 158 members. Furthermore the secretary reported a deficit of $292.50 in the operation of the association, which had total expenses that year of $3333.00. While it is difficult to explain this precipitous decline in membership, one reason may have been the fact that the dues were doubled in January 1911, from $5.00 to $10.00 a year.

Another reason for this decline in membership was due, in part, to the fact that the recruiting of new members was no longer seen as part of the secretary's job. At the 1910 semi-annual meeting, it was concluded that, rather than solicit members, the association should concentrate its efforts in providing services of benefit to retail lumbermen and the lumbermen would then seek to join.

In an effort to stem the membership decline, a number of steps were taken. The first was the creation of *The Plan*, the association's first real continuing publication. The development of this magazine is dealt with in greater detail later. However, suffice it to say here that in presenting the proposal to the Board of Directors for approval, there were several motivations given: first, it would be an effective and attractive means of communication with members and would encourage membership; more importantly, *The Plan* was also seen as a profit-making operation that would help alleviate cash flow problems.

A number of other steps were also taken to stop the decline in membership. One was to consolidate and improve the grass-roots organization, and the Constitution and Bylaws were amended in 1913 to provide for subordinate exchanges to be affiliated with Pennsylvania Lumbermen's. This meant that local groups, with five or more members, could join the association as a group. It was felt that this would improve membership as association members in a regional or city exchange would encourage non-members to join. Eventually, this would lead to regional representation on the Board of Directors. Efforts along this line would begin at the end of the decade, and would officially be incorporated into the Bylaws in 1922.

More importantly, however, the association began to aggressively recruit members again. To do this, they enlisted an organization already

Advertisement from 1895–96 "Constitution and By-Laws."
Photo courtesy of EBMDA

Listing of officers of the PLPA, 1895–96 "Constitution and By-Laws." Photo courtesy of EBMDA

in existence, the Eastern Lumber Salesmen Association. At the April 27, 1915 Executive Committee meeting it was agreed

> That any member of the Board of Directors or the Membership Committee or the Secretary shall have authority to offer a commission of $5.00 to any lumber salesman who in the judgment of the Directors, Membership Committee or the Secretary, may be a proper person to solicit membership; the fee to be paid by order drawn on the Treasurer after the applicant has been elected to membership and his first year's dues are paid; and further resolved, after the application has been received and approved by the Secretary, said application shall be published in the next succeeding issue of *The Plan*.

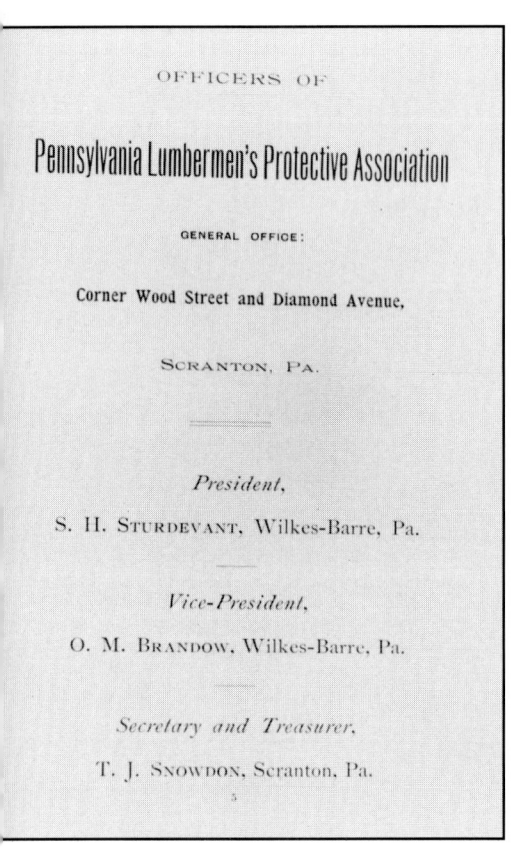

Within the year the Association began a massive letter-writing campaign to attract members. At a special meeting of the Board of Directors called to deal with the declining membership problem, it was agreed that the secretary was to compile a list of all retail dealers in the state by county. That list was to be sent to each county representative on the Membership Committee for them to check; subsequently a letter was to be sent to each of the retailers trying to get them to join. Rather than scatter the efforts throughout the state, the effort was initially centered on the region of the state considered to have the strongest support for the association: Dauphin, York, Lebanon, and Berks counties. By March 15, 1916, the secretary reported that he had sent out 600 to 700 letters. It was also agreed to provide up to $500.00 for the Membership Committee to use for complementary subscriptions for *The Plan* to be provided to non-members as an added inducement.

Also, because of declining membership, the association began to look beyond the borders of Pennsylvania for members. In 1914, at the semi-annual meeting, the Constitution and Bylaws were changed allowing membership by retail lumbermen outside of Pennsylvania, and immediately eight Delaware yards were admitted to membership. When the letter-writing campaign to attract new members was undertaken in 1915, it was agreed by the Board of Directors to also actively solicit members in New Jersey and Delaware.

Another endeavor to make the association more attractive was a proposal to create a "Junior Division" for the sons of members. Whether this actually occurred or whether it was simply an effort to provide activities at meetings that would be of more interest to a younger generation is impossible to determine from the official records.

As the decade progressed, the Pennsylvania Lumberman's Association, as did all trade associations, came under attack as organizations that restrained trade. In the case of *United States v. Eastern States Retail Lumber Dealers' Association* (1914), the court held, based on circumstantial evidence, that a conspiracy to fix prices existed because of discussions at trade association meetings. As a result, the semi-annual meeting removed from the Constitution the article dealing with irregular sales because of its possible illegality. It also created a Committee on Arbitration in an effort to avoid litigation in the future when disputes arose over improper sales.

The president of the association, Albert J. Thompson, addressed what he saw as the public's misunderstanding of their activities when he said, "we should let the consumer know that when he is doing business with a member of the Association, that he is not at the mercy of a combination formed for the purpose of extortion but one who appreciates that his highest duty is to give service to his community. That we are not members of a recently born class known as profiteers that the government has accused of taking too large profits and in this way prevents playing into the hands of mail-order concerns."

To address this growing concern against trade associations, the Pennsylvania Lumberman's Association adopted a code of ethics at its semi-annual meeting in July 1917. The code said:

> The Pennsylvania Lumberman's Association having for its object and purpose, among other things, the earnest desire to encourage and promote the practice of good ethics throughout the lumber industry and with that purpose and determination steadily before it, we feel it necessary to set forth those ethical principles which should serve as a guide in our business relations.
>
> 1. No order, once given and accepted, should be cancelled by either the buyer or seller, unless mutually agreed to by all the parties in interest.
>
> 2. All material upon which claims are made shall be kept strictly intact until the seller shall have had a reasonable time for inspection.
>
> 3. No combination of prices is sanctioned by our Association as we believe that wholesome, fair competition is essential to the life and success of any industry.
>
> 4. We condemn, however, as unsound and dangerous, any business practice which does not show a net profit, for if continued it must result in inexcusable failure or dishonest success.

Invoice from Smick and Harris, 1911. Photo courtesy of Smick Lumber and Building Materials Center

Forty-foot Hemlock log on way to mill, 1892.
Photo courtesy of Tinsman Brothers, Inc.

Circa 1925 old yard of Mizell Lumber and Hardware. After a flood in 1930, the yard was moved to its present location.
Photo courtesy of Mizell Lumber and Hardware Company, Inc.

5. In all our business transactions we shall accord such treatment to others as we in all fairness and honesty would have them render us.

6. Any violation of the above principles may be reported to the Board of Directors, and the Board is hereby empowered to act upon the same.

World War I had tremendous impact upon the association. As the country mobilized for war, transportation became a major issue and the association repeatedly petitioned the federal government concerning the lack of priority given the lumber industry on railroads as well as about the scarcity of building materials. These wartime transportation problems also added urgency to another issue that the association had been addressing in the past. As a result, during this period of time, PLA aggressively endorsed the efforts of the Atlantic Deeper Waterways Association, which advocated dredging the rivers of the East Coast to make them navigable for larger ships. The association also went on record calling for the nationalization of the Chesapeake and Delaware Canal.

The members of the association felt themselves hemmed in on all sides. On the one hand, the building industry was under constraints because of wartime scarcities and controls on transportation. On the other, they were beginning to feel the competition of mail-order companies, such as Sears-Roebuck, selling pre-cut house kits that were delivered to building sites. At the 1919 annual meeting, they applauded the actions of the Butterick Publishing Company in eliminating mail-order advertising from its publication and commended Butterick for its "Buy at Home" campaign.

Additionally, associations such as PLA were under legal challenge. Yet, as good citizens, they were being asked to help pay the cost of war through new taxes such as the income tax and war savings plans. This frustration was captured by President E. K. Moyer, who wrote in his annual report to the association:

> The year 1918 will go down in history as recording the greatest and most important events since the birth of Christ. For a year the country has been turned topsy-turvy and the greatest skill and effort was required to bring it to a successful conclusion; business was not sure

"At any rate, we won't have any difficulty finding this place," remarked Johnny.

"It's as simple to erect one of our houses as for a kid to make building blocks."

From early in the twentieth century, the association fought firms such as Sears-Roebuck that sold mail-order houses. These two cartoons attacking the practice are from The Plan, *May 1927.*
Photo courtesy of EBMDA

from one day to the next; the building material man was between two fires—embargoes on lumber ordered and the ban on new buildings; his business was restricted yet he was repeatedly called upon to show his patriotism by buying bonds, war saving stamps, etc.

The association, itself, attempted to deal with these changing times in a number of ways. First, it continued its recruitment efforts and especially focused its energy on recruiting new members from Delaware and South Jersey. One way they recruited in these areas was to hold annual or semi-annual meetings in Wilmington and at the Jersey Shore, in such locations as Ocean City and Cape May. They also attempted to revive contacts with the Western Pennsylvania Retailers Association, and quarterly meetings were held with the officers of each organization as a way of developing closer cooperation. To increase contact among the members of the two groups, the members of the Western Pennsylvania group were invited to subscribe to *The Plan* and news items about their organization were included in the magazine.

Advertisement for circular saw mill, circa 1900.
Photo courtesy of EBMDA

Additionally, they rewrote the Constitution and Bylaws in an effort to bring them into legal conformity with changes in the business environment. Gone were the restrictions on activities. Instead, they incorporated the code of ethics enacted earlier and statements opposing any actions that might stifle competition, limit production, restrain trade or regulate prices.

In an effort to stimulate the building material industry, the association became very active in encouraging the establishment of building and loan associations. In fact, the association provided free of charge to members "all information and assistance necessary to the organization and incorporation of building and loan associations and to encourage through the medium of *The Plan*, the organization of such an association, as a stimulus to the lumber business." The association also actively supported a national campaign entitled "Own Your Own Home."

At the end of the decade, the association went on record supporting attempts to impose national standardization in the manufacture of one-inch lumber and endorsed having it manufactured to thirteen-sixteenth.

They also increased the dues by 50 percent by raising the fee from $10.00 to $15.00, and increased the salary of the secretary of the association to $75.00 a month. The treasurer would report at the 1920 annual meeting that the organization had once again returned to fiscal soundness with total annual expenses for the year amounting to $5,418.19 and cash balance of $1,035.55.

1921-1930

It is difficult to provide a detailed history of this period of time because of a gap in the records from 1924-1934. As a result we only have information for the early period of the decade.

It is obvious, however, that a number of changes were being undertaken as the association began to adapt to the postwar world. Initially, after many years of debate, the territory of the association was divided into groups of local units and each local unit was represented on the Board of Directors. This reorganization was approved at the 1922 annual meeting with local retail lumber dealer's associations being represented on the Board of Directors if 50 percent of the local association belonged to PLA.

*Circa 1900 picture of early office of Eisenhauer, MacLea and Company, 318 West Fall Avenue, Baltimore, Maryland. Mr. Eisenhauer is in the doorway, Mr. MacLea is in the driveway.
Photo courtesy of EBMDA*

At the same meeting, an effort was defeated to change the name of the organization to "Eastern Pennsylvania Lumber Dealers Association." Instead, a resolution passed calling for the name to be changed to "The Pennsylvania Lumbermen's Association and shall cover the territory of Eastern Pennsylvania, Delaware and Southern New Jersey."

The 1922 meeting also created a graduated dues schedule based on volume of business:

Up to $50,000	$15.00
$50,001-$100,000	$20.00
$100,001-$150,000	$30.00
$150,000-$200,000	$35.00
$201,000-$300,000	$50.00
$301,000-$400,000	$65.00

J. Gibson McIlvain and Company, one of the oldest lumber dealers in Philadelphia, Pennsylvania, circa 1920. Photo courtesy of Pennsylvania Lumber Museum

$401,000-$500,000 $80.00
over $500,000 $100.00.
FOR COMPARISON:
The current dues structure of the Association in 1992 is:

Sales volume up to $1,000,000	$260
$1,000,001 to $1,500,000	$335
$1,500,001 to $2,500,000	$385
$2,500,001 to $3,500,000	$435
$3,500,001 to $5,000,000	$485
$5,000,001 to $7,500,000	$535
Over $7,500,000	$585

This dues structure has been changed only once since 1969.

The 1922 meeting also approved a Bylaws change calling for compulsory arbitration between members of the association and the Wholesale Dealers Association. Once again, it would appear that efforts to provide a legal way of dealing with wholesalers selling to consumers was proving difficult to establish and this was only the latest effort.

The early 1920s also saw the PLA becoming more involved nationally. First, in 1921 they joined the U.S. Chamber of Commerce. Probably most important, however, was the decision in April 1922, to join the National Retail Lumber Dealers Association. While PLA had generally supported the actions of the association, this would begin an official association with the national organization that would become stormy and contentious at times, as we shall see.

Another action taken in the postwar period that would set a pattern for the future was the decision to sell uniform order blanks to members. Thus began a practice of having the association provide a large number of services for members where savings based on economies of scale could be passed on to the members by bulk ordering. In this particular case, members could obtain triplicate set order blanks for $.80 per 100 in sets of 500 or more. In 1923, the association provided a cost accounting system for members to use in their operations. In 1924 beginning discussions were held about creating an Association Credit Bureau with other lumber associations, particularly the Philadelphia Lumber Exchange.

The association also created the position of field secretary (assistant secretary) for the first time and appointed J. H. Reiter, a student at Haverford College, to the position at a salary of $1,500.00 per year. Reiter was to divide his time between *The Plan* and membership recruitment. As a result, half of his salary was to be paid by *The Plan* and he was to receive an incentive payment of "25 percent of the first year's dues of all new members he should procure, to him not less than $5.00 for each member." He also received an expense account of $100.00. In addition to his editorial work and advertising solicitation with *The Plan*, Reiter concentrated on organizing local units as constituent groups within the association. This arrangement lasted only two years, however. It became obvious that the job was too big and that if Reiter spent his energies on *The Plan*, then recruitment suffered, and, if he traveled to build up the membership, then *The Plan* suffered. As a result, Reiter resigned in 1924 after it was decided to split the job into two, with each position having a lower salary.

Efforts were also undertaken to consolidate the offices of the lumber-related associations in Philadelphia. When questions arose over the excess capital being held by the Pennsylvania Lumbermen's Mutual Fire Insurance Company, a committee was created to investigate. It reported to the association that rather than reduce premiums, as had been suggested, they should invest their capital in a building that would house the

Wood piled on a pier in Northeast Philadelphia, circa 1920.
Photo courtesy of Pennsylvania Lumber Museum

Hemlock log rafts on Delaware River from Downsville, New York, at landing of Tinsman Brothers, Inc., circa 1892. The rafts are 60 by 200 feet.
Photo courtesy of Tinsman Brothers, Inc.

lumber industries around Philadelphia. However, when prospective tenants committed only to two floors of the new building, the plan was dropped.

At their suggestion, the PLA held periodic discussions concerning consolidation with the Building Material Dealers Association of Eastern Pennsylvania. While a number of meetings were held, the plans were shelved in 1923 as being "impractical."

While it is difficult to give details about activities at the end of the decade, because of lack of records, it is obvious from what few details remain that the association was on the verge of a radical transformation of its purpose and activities. Increasingly, the Pennsylvania Lumbermen's Association was moving away from the activities that had initially motivated its founding; actions that the government would eventually declare illegal; or activities in restraint of trade. The association was moving more to what might be called the "modern" approach to activities for such associations—providing a variety of services to retail dealers as a group that would not be available to individual dealers. This would include such things as new techniques in merchandising of modern building materials; model store displays and building construction; educational programs related to specifics of the retail lumber trade, as well as programs providing education on modern business practices; insurance programs; updates on governmental regulations and legislative activities, as well as services such as cooperative purchasing of low demand retail items; and items necessary for the conduct of business.

These were all activities that the Pennsylvania Lumbermen's Association had begun to do early in the decade, but which were accelerated by the onset of the Depression and the coming of the New Deal with all its federal regulations.

Yard scene, circa 1900–1905, showing Walter Smith of Tinsman Brothers, Inc. Photo courtesy of Tinsman Brothers, Inc.

A yard worker at Ohio Valley Lumber Company, circa 1900.
Photo courtesy of Ohio Valley Lumber Company

Load of shingles being delivered by Walter Smith, circa 1890–95, Tinsman Brothers, Inc.
Photo courtesy of Tinsman Brothers, Inc.

Eisenhauer, Mac Lea Company, Baltimore, Maryland, circa 1910. Photo courtesy of EBMDA

Cartoon from The Plan, *October 1926, showing the evils of price cutting. Photo courtesy of EBMDA*

Photo courtesy of J. D. Loizeaux Lumber

Galliher & Huguely, Inc.

LUMBER
SHERMAN AVE. & W. N.W.
PHONES NORTH 486-487

WASHINGTON, D.C. May 29, 1923

SOLD TO **P. G. Whitty Co**
1301 — Mass Ave NW

THESE PRICES ARE NET. NOT SUBJECT TO DISCOUNT

DATE	TICKET NO.			
May 6	11325	2 7/8 x 7/6 — 1 1/8 Drain Boards 2¢ ea		4 00
7	11350	3 PC 7/8 x 12-14 — W.P. Case 42. 46 00		1 93
8	11477	1 PC 1 3/8 Cove 10 —	4¢ L	40
		1 v 1/2 x 12-4 C W Pine	12¢	48
15	11555	4 PC Rock Lath	25¢ ea	1 00
		1 Box Sheet Rock Filler	50¢ each	50
26	12533	8 Lin Ft 1 3/8 Cove	2¢ L	16
				8 47
				08

6/11/31

Photo courtesy of Galliher & Huguely, Inc.

The H. G. Green Lumber and Coal Company around the turn of the century. As shown above (C), coal was brought to the yard in barges on the Mantua Creek, Mt. Royal, New Jersey.
Photo courtesy of H. G. Green Lumber Company

Window seal for participating dealers from "Portfolio of Promotion Campaign for Certified Lumber Standards," a campaign initiated by MALA circa 1940. Photo courtesy of EBMDA

Chapter III
The Middle Years: Middle Atlantic Lumbermens Association

In 1933 the name of the Pennsylvania Lumbermen's Association was changed to the Middle Atlantic Lumbermens Association (MALA). After a decade of discussion, it was felt that the new name more accurately reflected the scope and geographical impact of the association. Since, by the time of the name change, the association already had member dealers outside of Pennsylvania, more aggressive efforts were made to recruit dealers in Delaware and Maryland, particularly around Baltimore, the District of Columbia, and Western Pennsylvania. Again, because of the lack of records for the period 1924–1934, it is impossible to track the detailed discussions that took place. However, we know from what survives that this name change debate had been going on for more than a decade. As early as 1922, a name change had been proposed but was turned down by the traditionalists. In all likelihood, it was the drain on the membership associated with the Depression and the collapse of regional associations in these other areas that finally forced them to allow for the more geographically inclusive name to be substituted.

Also, because of the lack of records, it is impossible to accurately portray the impact the Depression had on the association. What is obvious, however, is that the coming of the New Deal reoriented the activities

Page from The Plan, *March 1931, highlighting "Build Now" campaign as a way of fighting the Depression. Photo courtesy of EBMDA*

Photo from The Plan, *March 1931, emphasizing salesmanship. Photo courtesy of EBMDA*

"Blue Eagle," symbol of the National Recovery Administration. By displaying the Blue Eagle, retailers, indicated their support for and compliance with the "Codes of Fair Competition" negotiated by the NRA. Photo courtesy of National Archives

of the association and took many of the practices that were suspect before the Depression, such as price regulation, and gave them official imprimatur under the National Recovery Administration (NRA). In particular, the industrial code enforcement provisions were a part of this controversial legislation.

NRA had two purposes. The first was to stabilize business with codes of fair competitive practices, and second, to generate more purchasing power by providing more jobs, raising wages, and defining labor standards. Within each industry, committees representing management, labor, and government drew up the codes of fair practice. Various trade associations, such as MALA, became vital players in the game of industrial code making and enforcement. Since the lumber industry was one of the earliest codes to be arranged in 1933, the entire industry was under the Blue Eagle symbol of NRA as it attempted to stabilize prices and curb the economic slide from the Depression through price and cost fixing schemes. The lumber industry also supported uniform wage and hour provisions that were agreed to by the government, industry, and the individual retailers. Under these provisions, the government delegated the power to devise and enforce controls to industrial leaders and trade associations such as MALA. As a result, the power of trade associations was greatly enhanced. To illustrate this point we need only quote Hugh Johnson, the head of the National Recovery Administration who, speaking before the National Association of Manufacturers, stated: "NRA is exactly what industry organized in trade association makes it . . . before NRA, the trade associations had about as much effectiveness as an Old Ladies' Knitting Society; now I am talking to a cluster of formerly emasculated trade associations about a law which proposes for the first time to give them potency." Under NRA, MALA became one of the regional enforcers of the code provisions for the retail lumber industry, and it was paid a fee for its efforts. All of the retail lumber dealers who wished to be a part of NRA and display the Blue Eagle, showing compliance with the provisions of the Code, paid a fee to the association.

These actions were not without controversy, not only within MALA itself but also within the country. By 1935 there was disagreement over the codes themselves and about how they were being enforced. The MALA Board of Directors decided to withhold further payment of their share of the expenses until the uniform enforcement of the codes could be agreed upon. This never happened and, in fact, NRA was declared unconstitutional by the U.S. Supreme Court in 1935. By that time, MALA, as well as other similar retail trade associations, had begun to emphasize anti-chain store and "fair trade practices" legislation in various state legislatures. While the passing of NRA was not mourned by many, it left its mark on the trade associations as well as the country. With dramatic suddenness, the codes had set new standards such as the forty-hour workweek and the end of child labor, and it was difficult to retreat. At the same time, the codes advanced the trend toward stabilization and rationalization that were becoming the standard practice of large and small business and which had been advocated by most trade associations throughout the 1920s.

This flirting with statism was a mixed blessing for MALA. On the one hand, it invigorated the association and provided an infusion of cash at a time when dues from members were on the decline. However, it placed increased demands on the association that were difficult to finance given

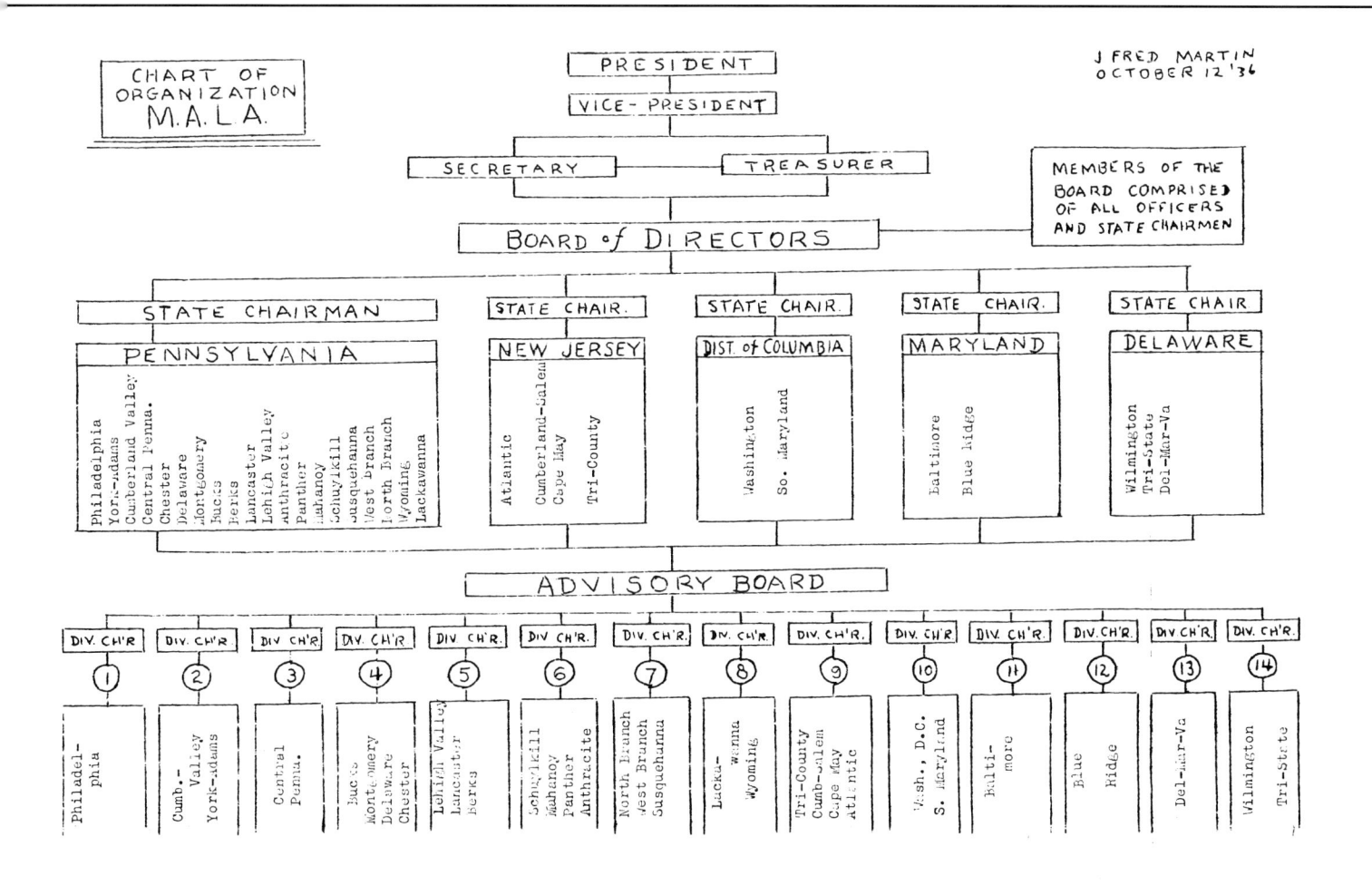

Photo courtesy of EBMDA

Chart from the Minute Book, Board of Directors, MALA, in the hand of J. Frederick Martin, showing the newly established organizational structure of MALA in 1936. The divisional chairmen and advisory board structure would not work out; it would be replaced shortly thereafter.

the current membership and the dues collected. In 1935 it was reported that it had 250 dues paying members. However, at the same meeting of the Board of Directors, it was reported that when the NRA was first implemented, 830 retail lumber dealers had paid the initial code assessment. Thus, association membership accounted for only about a third of the eligible retail dealers in the region covered by MALA.

As the code enforcement provisions of NRA waned, MALA began to provide a variety of services to members. One of the first initiatives was to create "The Lumbermen's Purchasing Corporation" (LPC), which was to act as a "spearhead" for MALA. This new corporation was to act as a clearinghouse and receive all orders from members and place orders with shippers. It would keep members informed of market conditions in the lumber and building materials market; coordinate cooperative purchasing among the various local units; and arrange most favorable terms with manufacturers and wholesalers. Each member dealer would place his business just as he had in the past, but through LPC instead of directly with the shipper. This would allow the corporation to check on demand, volume of purchases, shippers preferred by each dealer; would keep members posted on market prices, favorable buying opportunities, special prices for bulk shipments and would secure the lowest possible prices for members without disturbing the current distribution channels. The Board of Directors also began looking into a cooperative purchasing arrangement for slow-moving commodities.

While this purchasing corporation sounded like a good idea to the members of MALA, the wholesalers, manufacturers, and jobbers began to complain bitterly of the arrangement and threatened retaliation by selling directly to contractors. The association's Board of Directors was divided on the issue with some believing that it was only through such activities that individual retailers could counter the activities of chain stores such as Sears-Roebuck and others. MALA retreated from their proposal, however, and backed away from the corporation. The Lumbermen's Merchandising Corporation, however, went on to be a very successful operation under the leadership of former MALA official, James Buckley.

They moved forward on a number of other fronts to provide needed services to members. In 1935, they created an Inspection Service that provided inspectors on call who would provide inspection services and

Photo courtesy of EBMDA

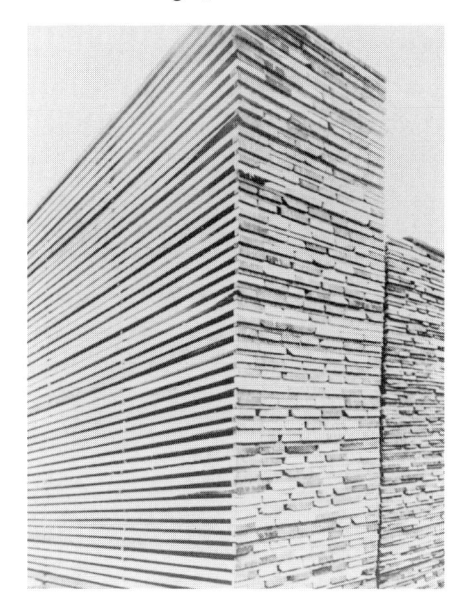

Photo courtesy of EBMDA

tally information on any lumber purchased. At the same meeting, the Board of Directors agreed to establish a service to supply what amounted to a legislative information bureau to members. This new service would provide members with information on all legislation passed by the state and federal governments that affected business interests and would provide summaries for inclusion in *The Plan*. Additionally, the association arranged for Wolf and Company, an accounting firm, to supply representatives throughout MALA's service area to assist members in income tax preparation, social security forms, and the capital stock tax.

All of these services were additions to the already long list of services provided by MALA such as the credit services, arbitration, market letters, and merchandising counseling. This meant a substantial increase in costs that had to be covered by increased revenues — dues. At the same time that the association was attempting to deal with these pressures, the Commissioner of Internal Revenue informed MALA that they were subject to federal income tax because of their operation of the architectural service and *The Plan*. The association agreed to pay the back taxes and began to look into mechanisms of spinning off

Photo courtesy of EBMDA

View of the yard and truck of Brosius and Smedley, 1930, Wilmington, Delaware.
Photo courtesy of Brosius-Eliason, Inc.

Brosius-Eliason's new truck, 1930.
Photo courtesy of Brosius-Eliason Inc.

Workers stacking lumber, Ohio Valley Lumber Company.
Photo courtesy of Ohio Valley Lumber Company

these activities or creating separate corporations so that MALA could return to a nontaxable entity.

These new activities placed a great strain on the finances of the association at a time when payments from the code enforcement were rapidly declining. By 1935, only 400 dealers were paying their assessments, whereas 830 had paid in 1933. At the same time, membership was also declining for two reasons: (1) business was so bad that many of the smaller dealers could not even afford the minimum membership fee, and (2) much of the real work of the association was lost under the administration of the NRA codes. Dealers saw no need to belong to MALA while the codes were being enforced. Consequently, the association was experiencing one of its periodic cash flow problems generated by low membership. As a result, in 1935 a complete review of the association was begun, not only in terms of structure and operation, but also in terms of the dues structure and method of payment. For 1935, the board established a dues payment of 1/4 of 1 percent of sales payable monthly.

The new dues structure drew immediate and widespread opposition, and a significant number of members refused to join. Thus, almost immediately the MALA board began to rethink the entire dues structure as well as the organizational structure of the association. The result of all of this was a period of chaos and turmoil for the association. However, within a year the structure had been reorganized and a new dues structure was in place.

At this point, it might be interesting to set forth the actual budget for MALA for 1936 to show how the dues were being spent:

Expenditures:	
Executive Salaries	$5,000.00
Office Salaries	3,000.00
Rent	1,200.00
Telephone & Telegraph	300.00
Travel Expenses	1,140.00
Postage, Stationery & Office Supplies	900.00
Office Maintenance	36.00
Dues	600.00
Miscellaneous	720.00
Accounting fees	960.00
Total	$13,856.00

Income:	
Dues	$7,000.00
Plan Advertising	3,000.00

Architectural Services	2,000.00
Income from Investment	520.00
Annual Meeting	400.00
Total	$12,920.00

FOR COMPARISON:

The combined budget of the Association and MALA, Inc., fifty years later, in 1986, was a hundred times greater. Some comparable line items to 1936 are as follows:

Expenditures:

Executive Salaries/Bonus	$170,484
Office Salaries	267,441
Payroll taxes	34,867
Convention	105,658
Dealer directory/buyers' guide	21,963
Forms program	347,524
Rent	11,580
Meeting expense	42,478
Telephone	13,900
Travel Expense	64,850
Postage	25,373
Stationery & Supplies	7,336
Office Maintenance	27,285
Dues	19,857
Miscellaneous	10,780
Professional fees	10,213
Education & training	23,697
Leasehold expenses	24,324
Comparable Total	$1,229,540

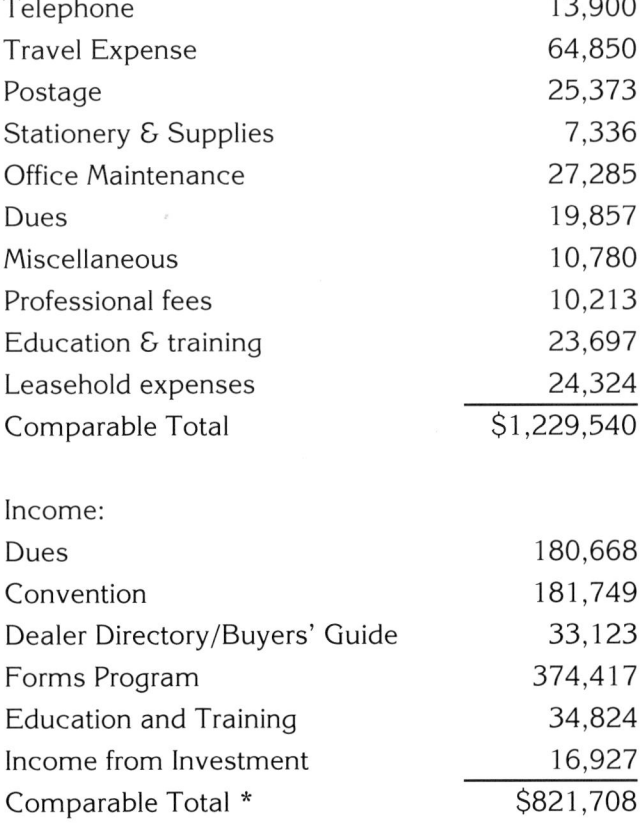

View of mill work shop, early 1930s.
Photo courtesy of Nelco Lumber and Home Center

Income:

Dues	180,668
Convention	181,749
Dealer Directory/Buyers' Guide	33,123
Forms Program	374,417
Education and Training	34,824
Income from Investment	16,927
Comparable Total *	$821,708

* Additional income items produced a balanced budget

    Under the old organizational structure, MALA had thirty-one local units, each electing a member to sit on the Board of Directors. In addition,

MEMBERSHIP ROSTER
and
GENERAL DIRECTORY
of
Retail Lumber Dealers
in Eastern Pennsylvania, Southern New Jersey,
Delaware, Maryland and District of Columbia

corrected to
SEPTEMBER 1, 1934

Price, $1.00

MIDDLE ATLANTIC LUMBERMENS ASSOCIATION, INC.
1102 Girard Trust Building,
Philadelphia, Penna.
RITtenhouse 7892

Photo courtesy of EBMDA

there were nine directors at large, bringing the total to forty directors and eight members of the Executive Committee. When a combined meeting was held, forty-eight people could potentially attend. This was felt to be too unwieldy.

The new structure provided for fourteen separate divisions, each embracing one or more local units. Each local unit would continue to function as it had in the past and would hold its own meetings. Each of the fourteen divisions was headed by a chairman who would call quarterly meetings of the units, or as often as desired, in order to consider and discuss local divisional problems. Annually, at a designated divisional meeting, the officers of each local unit, comprised in the division, elected a division chairman who presided over divisional meetings and acted as a contact man between his group and MALA. The divisional chairmen constituted an Advisory Board to the regular Board of Directors of the association. The Advisory Board was to meet with the regular Board of Directors as frequently as necessary.

Each state subdivision of the association was to have a state chairman who was to be elected at the Annual Meeting of his group. The state chairmen, consisting of five in all, constituted the Board of Directors of MALA with the president, vice-president, secretary, and treasurer, acting as directors ex-officio.

The president, vice-president, secretary, and treasurer were to be elected at the annual meeting. The secretary and treasurer did not have to be dealer members. This new structure was approved and implemented in 1936. However, in 1938, the structure was modified to have five directors from Pennsylvania, one from Delaware, three from Maryland, one from the District of Columbia, and two from New Jersey.

A new dues structure was implemented in 1936 that called for a minimum payment of $25 a year for any volume of business up to $50,000 and a sliding dues scale for member dealers with business volume above that amount. In 1938, the minimum payment was reduced to $15. Also allowed was quarterly payment of dues in order to soften the burden of payment.

With these organizational changes behind them, MALA then began on a long-term membership drive that would bring back many of the delinquent members and add new members to the rolls. Part of that membership drive was the hiring of a field secretary, Robert A. Jones, who was to assist with the publishing of *The Plan*, solicit advertising, and recruit new members. In June 1936, the association had only 153 dues-paying members from a potential pool of approximately 800 retail lumber dealers. By the end of 1940 they had 340 members. Membership was encouraged by having members of the Board of Directors call on non-members within their areas. It was also encouraged by providing incentive pay to employees who brought in new members. For example, in 1938 it was agreed that if the secretary, Fred Martin, and the new field secretary, Bob Jones, were successful in increasing membership, any surplus money in excess of $2,000 held in association accounts after all expenses had been paid would be split into thirds: one-third being given to Martin; one-third to Jones; and one-third to the association. This arrangement apparently worked for the benefit of all concerned, because the same incentive system was renewed in 1939.

During the rest of the prewar decade, the association continued to push for fair trade practice bills in state legislatures and even attempted to have Governor Earle of Pennsylvania amend a call for a special session of the legislature to deal with unfair trade practices legislation. However, efforts to push for fair trade practices legislation stopped by the end of the decade as the industry came under scrutiny by Thurman Arnold at the Department of Justice. At this time the association began to lobby state legislatures more aggressively for specific pieces of legislation as they affected the retail lumber trade. For example, MALA retained the firm of Thompson and Baird, Esq, "who have excellent political connections," to lobby the Pennsylvania legislature for exemption in case of emergency and for night watchmen from the forty-four-hour-a-week labor law.

MALA also continued expanding member services such as hiring a traffic manager to assist members negotiating with railroads to reduce rates on lumber going from Philadelphia to surrounding areas. The association also began a dispute with the national association which, during the next decade, caused a major split between the two organizations. The disagreement was two-fold: membership fees and the proper role of the national association. MALA's membership fee was assessed at $1,000. The Association said it would pay $2.00 per member and no more.

As the decade closed, the Board of Directors took two actions that would set a pattern for the coming decades. First, they began an insurance service for the association. In October 1938, the Board of Directors

MALA ad from The Plan, *September 1937. Photo courtesy of EBMDA*

By-Laws

of the

Middle Atlantic Lumbermens Association

Headquarters
1102 Girard Trust Building
PHILADELPHIA, PA.

Photo courtesy of EBMDA

began negotiations with an outside agency to provide a variety of insurance services to member dealers beyond the longstanding arrangement that they had with Pennsylvania Lumbermen's Mutual Fire Insurance Company. Secondly, they decided to "commercialize as many services as possible." As a result, Fred Martin and Robert Jones were directed to obtain insurance broker's licenses and to collect a brokers fee to help cover the cost of their salary. Within six months they reported that they had delivered seven policies and had filed as broker of record in sixty-two instances. In 1940 the association treasurer, Ray Latshaw, reported that forty-four policies had been written and commissions in excess of $2,300 were paid to the association.

THE WAR AND POSTWAR PERIOD

The year 1941 saw another reorganization of the association. Of major importance was the fact that J. Frederick Martin, who had served as the chief staff officer since 1910, began to withdraw from the day-to-day operation. Robert A. Jones replaced him and was named manager; he would eventually be named executive director and then executive vice-president. The Bylaws were rewritten and the Board of Directors was reorganized with twenty members representing twenty geographical districts. The Districts were as follows:

1-Philadelphia County
2-Montgomery and Bucks Counties
3-Northampton, Lehigh, Carbon and Monroe Counties
4-Pike, Wayne, Lackawanna and Susquehanna Counties
5-Luzerne and Wyoming Counties
6-Sullivan, Lycoming and Clinton Counties
7-Snyder, Union, Northumberland, Columbia, and Montour Counties
8-Schuylkill County
9-Berks and Lebanon Counties
10-Chester and Delaware Counties
11-Lancaster County
12-York and Adams Counties
13-Cumberland, Perry and Dauphin Counties
14-Franklin County, Pa., and Washington County, Md.
15-Frederick, Carroll, N. Baltimore, Harford, Howard, and Montgomery Counties
16-Baltimore, Anne Arundel, Prince George, Calvert, Charles, St. Marys, and Washington, D.C.

17-Cecil, Kent, Queen Annes, Talbot, Caroline, Dorchester, Wicomico, Somerset, Worcester, Md.
18-Newcastle, Kent and Sussex Counties, Del.
19-Salem, Cumberland, Atlantic and Cape May Counties, N.J.
20-Gloucester, Camden, Burlington, Counties, N.J.

Membership in the Association was open to "any individual, firm or corporation operating a business with the customary facilities and equipment for the assembling and storage of lumber, or of lumber and other materials or products customarily employed in building construction, and its sale at retail, for the purpose of making a profit upon such operations; that maintains a stock sufficiently varied and in sufficient quantities to care for all the reasonable demands of public service; that maintains an office with someone attendant thereto, the facilities of such office as well as those of the yard and its equipment being available for the transactions of business and the supplying of materials during all customary business times."

The reorganization of the association also called for the creation of a Finance Committee which began to do a careful review of the activities of the association in an effort to cut costs. One of the first things they did was to move the offices from the Girard Trust Building to 1528 Walnut Street.

There were two issues that would play a major role in the early part of the war decade. The first was the fact that, in 1938, the federal government had indicted all retail lumber dealer associations, including MALA and the National Retail Lumber Dealers Association, for restraining trade. In particular, the federal government charged that a lumber association in Colorado had organized a boycott against manufacturers of lumber who sold to other than lumber dealers in violation of the antitrust laws. In 1941 this would result in MALA becoming a party to the consent decree that settled the case. Additionally, the association had to provide $2,500 for payment for expenses associated with the settling of the case.

At the same time relations with the National Retail Lumber Dealers

Cover of flyer describing Certified Lumber Standards. It was a way of encouraging the use of local lumber dealers by contractors and architects to ensure quality through guaranteed use of recognized and approved standards for all lumber sold for construction.
Photo courtesy of EBMDA

What Lumber Means to the Boys at the Front

HOW many carloads of lumber does it take to pack a Liberty Ship and send it on its way overseas to the battle front? How many millions of board feet of lumber have been used to rebuild docking facilities at Pearl Harbor, in Africa, Italy and how much lumber will be needed on the European beaches of the Second Front? How big are the stock piles set aside in Allied countries to meet these urgent needs? The huge amount staggers the imagination and there's no immediate let-up in sight encouraging as the war news from overseas is these days.

Lumber may be short on the home front but it takes the boys on the fighting front to describe what shortages really mean. A Michigan lumber dealer recently received a letter from his son who is with the Seabees in the South Pacific. In commenting on the lumber situation in that far-flung sector, he said:

"Lumber is the most valuable of all supplies next to ammunition and rations here in the Southwest Pacific. Lumber and screen are the only things that keep us from having dysentery and malaria all the time. We beg, borrow, steal and fight for every little stick."

A short time before Ernie Pyle, famed war correspondent who is living with the boys at the front and describing their day-to-day activities as no one else has been able to do it, told an unforgettable story of the dugout homes in which many of them are living.

"The soldier's dugout," he said, "is made by digging a square or rectangular hole about shoulder deep, then roofing it with boards and logs, piling earth on top of that... Some dugouts have board walls to keep the sand from caving in. Others use the more primitive method of log supports in each corner with shelter halves stretched between them to hold back the sand .. It takes a lot of lumber to shore up all those thousands of dugouts. The boys rustle up anything they can find out of old buildings... The finest dugout I've seen belongs to four officers of a tank company. This dugout is as big as the average living-room back home. You can stand up in it and it has a rough wooden floor... They even have a big white dog, slightly shell-shocked, to lie on the hearth."

As empty as dad's lumber yard may appear, it would be a veritable gold mine to the lads in the armed forces who spend days looking for a small piece of wood to patch up a dugout where they hope to get a few minutes' sleep.

No advertising or promotion program is so big that it can offset the effects of discourtesy.

It used to be that retailers had many complaints on service but few on quality. Today in many industries the complaints on quality in some cases outrun the complaints on service ten to one.

I'm Sick

So you're sick of the way the country's run,
And you're sick of the way the rationing's done.
And you're sick of standing around in line
You're sick you say—well that's just fine.

So am I sick of the sun and heat,
And I'm sick of the mud and jungle flies,
And I'm sick of the stench when the night mists rise,
And I'm sick of the sirens wailing shriek
And I'm sick of the groans of the wounded and weak.

And I'm sick of the sound of the bomber's dive,
And I'm sick of seeing the dead alive,
I'm sick of the roar and noise and din,
I'm sick of the taste of food from tin.

And I'm sick of the slaughter, I'm sick to my soul,
I'm sick of playing a killer's role,
And I'm sick of blood, and of death and the smell,
And I'm even sick of myself as well.

But I'm sicker still of a tyrant's rule,
And conquered lands, where the wild beasts drool,
And I'm cured dam quick when I think of the day
When all this hell will be out of the way.

When none of this mess will have been in vain,
And the lights of the world will blaze again,
And things will be as they were before
And the kids will laugh in the streets once more.
And the Axis flags will be dipped and furled
And God looks down on a peaceful world.

—*Contributed by Legionnaire C. E. Daley*

A page from The Plan, *June 1944, relating lumber to the war effort.
Photo courtesy of EBMDA*

Association, which had been on shaky ground for over a decade, came to a climax and MALA resigned from the national organization. While part of this was due to issues relating to the federal indictment, there were a number of other festering issues. One was the manner in which the national association presented assessments against MALA without what MALA thought to be adequate time for comment and discussion. For example, the national association decided to appeal a federal decision under the Fair Labor Standards Act and presented a bill for $2,000 to MALA for their share of the legal fees. In addition, they proposed that MALA contribute $4,000 toward the legal defense of the federal indictment. In the opinion of the MALA board, the wage-hour decision should not have been contested and the NRLDA should have pled *nolo contendere*, as the other associations had done. The bottom line in this debate, however, was that MALA did not believe that the national association was adequately representing the retail lumbermen and they resigned. However, having made their point, as part of the arrangements for signing the consent decree, they withdrew MALA's resignation from NRLDA. In the future, these same issues would surface again and cause another split.

This concern over unexpected bills from the NRLDA and MALA's aggressive stance was not due to a precarious financial position. While there was a temporary shortfall in 1941 because of charges associated with the legal fees involved in the consent decree, the association was beginning to prosper, as was the economy. The membership drive had been successful; dues payments were up; and they were taking a more aggressive stand and beginning to provide a number of new services for members. By 1944 the executive director, Bob Jones, reported that "the general tone of the association is noticeably better than it has been in many years. We believe that when we protect one dealer we help build the reputation of all dealers."

Publications are dealt with elsewhere in this work. However, it should be mentioned here that during this period of time *The Plan* was reorganized and set up as a separate corporation. This was done, in part, to help settle continuing problems with the Internal Revenue Service over the tax status of the association. By separating *The Plan* from the association, MALA was better able to argue its case for tax exemption, which they did in January 1946, and MALA was recognized as a business league qualifying for exemption. The exemption was made retroactive in 1944. At the same time, *The Plan*'s purpose was broadened and an effort was made to turn it into a journal that would speak to the concerns of a wider audience and address national concerns. Beginning in 1943 the association began to distribute another publication, *The Dealers Directory and Buyers Guide*.

It was also during this period of time that MALA began to develop an

The trucks of Galliher & Huguely in front of the U.S. Capitol.
Photo courtesy of Galliher & Huguely, Inc.

Photo courtesy of Galliher & Huguely, Inc.

MIDDLE ATLANTIC LUMBERMENS ASSOCIATION

PLATFORM OF SERVICE FOR RETAIL LUMBER DEALERS

In recognition of my responsibility to my community, the welfare of my country, and full employment for its people, AS A RETAIL LUMBER AND BUILDING MATERIAL DEALER, within the scope of my ability and resources, it is my purpose:

(a) TO MAINTAIN a construction sales and service center with adequate inventories and displays of counter and warehouse merchandise concerned with construction;

(b) TO INTEGRATE the elements which make up construction service packages, such as land, materials, equipment, design, fabrication, finance and utilities, so that the public will be afforded an opportunity to buy under a centralized sales and service responsibility the complete building package, installed or erected, ready to use, at a pre-determined price.

(c) TO COORDINATE the services of architects, builders, sub-contractors, mechanics, building finance agencies, producers and realtors, so that the public will receive the latest authentic information and guidance from these associates, who will also provide the expert skill to design, assemble, install, fabricate and deliver packages of building service.

I FURTHER PLEDGE:

1. TO PROSPECTIVE HOME OWNERS; a complete service with that quality in design, materials, construction and financing which will assure the maximum in health, comfort, convenience, beauty, durability and low cost maintenance.

2. TO ALL PROPERTY OWNERS; a complete service on repair, maintenance, remodeling, improvements and additions, including the workmanship of specially trained mechanics.

3. TO THE FARMERS IN MY TRADING AREA; a complete farm building service which will bring modern comforts and conveniences to the farm home, will provide for adequate shelter and increased productivity of livestock, and will conserve produce through modern storage facilities.

4. TO CONTRACTORS, INDUSTRIAL AND COMMERCIAL BUYERS; a complete line of materials and services for repairs, construction, maintenance and industrial uses.

5. TO ARCHITECTS, CONTRACTORS, REALTORS AND FINANCING AGENCIES; a central sales headquarters with sales leadership, adequate sales promotion and year-round creative selling to the end that the maximum attainable volume of construction will be developed to employ their services.

6. TO MY MANUFACTURING AND WHOLESALE SUPPLIERS; Full cooperation in maintaining adequate inventories, suitable displays and effective merchandising to the end that my trading area shall provide them an adequate volume of annual consumption of their products.

7. TO MY EMPLOYEES; Good working conditions, steady employment, income incentives together with thorough training and ample opportunity for advancement and increased earnings.

8. *TO MY COUNTRY, THE UNITED STATES OF AMERICA*, to do my full part in the perpetuation of the freedom of business enterprise which has made America great—a nation of home owners.

Pledge of Service approved by MALA, 1945.
Photo courtesy of EBMDA

extensive dealer education program. Under the leadership of Bob Jones, the "10 Point Program" was developed by MALA in an effort to provide useful, continuing education to member dealers and their employees. This program was eventually adopted by the NRLDA. At the same time, the association set up their first thirty-day course for dealers and their personnel at the Pennsylvania State College, beginning April 28, 1947. There would be two classes of twenty-five, and the Board of Directors approved providing a dinner for the graduates. This program proved so successful that a second one was scheduled for November. Similar types of education programs continue to the present. Later in the decade, the association provided correspondence courses for members and their employees and, in the early 1950s, established what was called the Retail Lumber Institute and Top Management Workshops and the Sheldon School of Sales Training. These provided additional educational opportunities for member dealers and their employees.

The budget for 1943 grew to $32,000 and by 1944 stood at almost $60,000 with a capital reserve of $11,000. The disagreement with the National Retail Lumber Dealers Association was forgotten in the short run as the two associations raised a "war chest" to fight the imposition of price controls.

The year 1945 also marked an important beginning for MALA. At the July 19, 1945, meeting of the Executive Committee, it was agreed to put the law firm of Saul, Ewing, Remick and Harrison on retainer at the rate of $500 a year. This was to begin a relationship with the law firm that continues today, a span of almost fifty years. The firm's assignment was to provide assistance to member dealers who were accused of violations of the Office of Price Administration regulations.

As an indication of the level of activity that the association was involved in by 1946, President Watson Malone III, reported to the Board of Directors that the association sent out three bulletins per week to members; handled over three hundred incoming calls a month; the staff attended approximately four dealer meetings a week and gave more than sixty talks. In addition, more than a thousand people visited the association's offices.

The organization was once again reorganized in 1947 with the major change being in the composition of the Board of Directors, which was to consist of from twelve to twenty-five members.

> The number of directors for each state within the territory of the Association shall at all times be in that proportion which the number of active members in the respective states bears to the total active membership of the Association, provided however, there shall be at least one director from each such State and the District of Columbia.

THE PLAN

MANAGEMENT CHECKLIST for RETAIL LUMBER DEALERS

MIDDLE ATLANTIC LUMBERMENS ASSOCIATION, Inc.
PHILADELPHIA

This Management Checklist is the Latest and One of the Finest Services the Association Has Developed for Members. It is a Flexible Loose Leaf, Business Razor with a Keen Cutting Edge. Blades Will Be Supplied Regularly for Each Section.

NOVEMBER, 1945

BE SURE TO READ

N. R. L. D. A. ADOPTS PLEDGE OF SERVICE [Page 9]
DISTRIBUTORS RESPONSIBILITIES [Page 11]

Cover of The Plan, *November 1945, featuring the "Management Checklist for Retail Lumber Dealers." These types of "aides" were an integral part of MALA's service to members. Photo courtesy of EBMDA*

The definition of membership remained the same, but the "semiannual meeting" of the association was eliminated.

As the 1940s ended, MALA and the National Retail Lumber Dealers Association once again disagreed over dues payments. When NRLDA announced a dues increase without proper consultation, the MALA Board of Directors agreed to pay only half of the amount. MALA's position was not so much opposition to the dues increase as it was to the lack of oversight by the National Board of Directors. A meeting was scheduled and NRLDA agreed to changes in their by-laws, and the balance of the dues were paid. The issue would surface again, however.

Also, another recurring theme appeared in the early 1950s: dues payments began to slip and the association, once again, had to engage in a crisis membership drive to shore up their financial situation. Expenses were cut, investments were sold, and the directors agreed to forego reimbursement for attending their meetings.

Finally, in surveying the history of the postwar period of MALA's history, one of the most important events was the action taken in September 1948. It was agreed to accept the recommendation of the Pension Plan Committee to create what would become known as the Middle Atlantic Lumbermens Association Group Insurance Trust. This would create and develop a highly successful group insurance program and pension plan and is discussed elsewhere in this work.

Editorial

CREEPING PARALYSIS

EARLY this month, the OPA sent out urgent requests to all local offices and civilian ration boards, directing that they immediately find out the names and addresses of every retail lumberman in their area, together with what lines and quantities of merchandise or materials were handled by them. The local boards were instructed, moreover, to return this information within 48 hours!

Here is another typical example of the frantic efforts of OPA to prolong their jobs at any cost. It is another of many stupid blunders by a self-appointed Gestapo to harass the small businessman. Like most of their crackpot regulations, it shows how little OPA understands the practical side of business they seek to permanently control. The results of such an insane request could only be confusion, while the time limit set by the Little Red Tape Princes to obtain such a tremendous volume of misinformation is laughable.

The OPA, however, is no laughing matter. If ever there was an outfit set on a steady encroachment of American rights and principles, it is the OPA. Their wartime controls were obnoxious enough, not so much because people objected to the controls, but to the arrogant, brow-beating method of enforcement. Their present attempts to outlive their usefulness by becoming peacetime parasites on business is an ominous challenge to our American principles of spiritual, political and economic freedom.

It is high time we put a stop to this un-American foe *of* business before the creeping paralysis of further OPA regulations puts a stop *to* business. It's more important than ever to tell your Congressman about these things and urge him to correct this intolerable situation. Certainly, the OPA is no safeguard against inflation as it is being operated today—it is just a troublesome road block to business and prosperity.

Robert A. Jones
Editor

Editorial by Robert A. Jones in The Plan, *October 1945, attacking the continuation of the Office of Price Administration, a wartime price control authority.*
Photo courtesy of EBMDA

THE FIFTIES AND THE SIXTIES

Once again we come to a period of time where the official records of the association are missing and it is difficult to reconstruct its history.

Your combat days are not over!

GET BEHIND THE VICTORY LOAN!

There's plenty of action ahead for fast-thinking industrial leaders in putting over the new Victory Loan! Your Victory drive is important because:

EVERY VICTORY BOND HELPS TO

1. Bring our boys back to the America for which they were willing to give their lives!
2. Provide the finest of medical care for our wounded heroes!

BOOST THE NEW F. D. ROOSEVELT MEMORIAL $200 BOND!

Urge all your employees to buy this new Franklin Delano Roosevelt Memorial $200 Bond through your Payroll Savings Plan! At all times better than ready cash, Victory Bonds are industry's "Thanks" to our returning heroes!

START YOUR VICTORY DRIVE TODAY!

Every Victory Bond aids in assuring peacetime prosperity for our veterans, our nation, your employees—and your own industry!

The Treasury Department acknowledges with appreciation the publication of this message by

MIDDLE ATLANTIC LUMBERMENS ASSOCIATION

★ *This is an official U. S. Treasury advertisement prepared under the auspices of the Treasury Department and War Advertising Council* ★

From the scattered reminiscences of several of the former presidents of the association, it is obvious, however, that, during this period of time, the organization was personified in the executive director, Robert A. Jones. It was also a period time of great change for the industry and the association, and the eventual success of the organization was due, in large part, to how dealers and the association responded to these changes. Again, the leader in this regard was Bob Jones.

Joseph E. Haenn, Jr., former president/chairman of the board of MALA, remembers it this way:

> It was Bob Jones' organization. He ran it, did the planning for it, and represented it to the trade across the country. The staff, the directors, and I were extensions of Bob Jones, and sometimes frontmen for him. He shared glory with those around him, and would do things to promote those who worked for the association. Perhaps this sounds dumb today, but it was good for those of us on hand at that time. He guided and taught lumber dealers, especially, his board and officers. He enjoyed attention himself, and had great feeling for the association, and liked it when the members of his business family received attention also.
>
> As lumber dealers, we were going through big changes. From being oriented mainly toward the professional builder and industrial customer, we were learning to invite the consumer in as a customer. We were opening retail stores. Sometimes it was hard. In the era of the early 1960s, I remember Bob had worked with *Life* magazine on a project that sent a woman reporter as a customer to a number of dealers in Philadelphia and South Jersey. She reported a number of places where she did not feel welcome. One place found the dealer sitting behind a desk with his feet up, and proclaiming, "I don't sell to women customers!"
>
> At the same time, the products we were selling were

An ad from The Plan *for the "Instant Lumber Piece Price," a publication used throughout the industry as a means of calculating the price of individual pieces of lumber. The federal government would eventually assert that it was a subtle mechanism for setting prices—its use would be discontinued.*
Photo courtesy of EBMDA

Ad from The Plan, *October 1945.*
Photo courtesy of EBMDA

A common postwar scene, war surplus equipment being used by businesses as a means of dealing with equipment shortages resulting from the war effort.
Photo courtesy of Maryland Lumber Company

Maryland Lumber Company shop, post–World War II.
Photo courtesy of Maryland Lumber Company

also changing. Ceiling tile and plywood paneling come to mind. It had not been too long before then that plywood sheathing had been accepted by the FHA. This knocked out the 1x4 and 1x5 T&G subflooring that had been the standard for years. And when a few years later carpet had been accepted by the FHA, it killed 80 percent of the oak flooring business within a year's time.

The visible changes in the lumber dealer started to show up about 1958–59 and so had progressed past the toddler stage by 1969—and one of the main prodding forces had been Bob Jones. He had been the visionary, the idea man, the master merchant, the showman, and the teacher. I don't remember him as being strong on product knowledge, but he was a giant when it came to merchandising and management. He really cared for his dealers and wanted them to succeed as businessmen, and he wanted them to be nice human beings at the same time.

Joseph W. Brosius, former president, concurs with this analysis of Bob Jones and his influence on the association:

> He was full of energy, full of ideas and had a terrific memory for faces and names. . . . He became Mr. Middle Atlantic Lumbermens Association. He was untiring in his efforts to help the dealers in the lumber industry. I spent many nights (with Bob Jones) visiting with local lumber groups, helping them plan and urging them to be better merchandisers. . . . Bob was an exceptional man. He was a religious man; he was very active in his community and was always willing to help. One of the problems he had with the association members was that they would call up in the morning and say "Bob, I'm coming to Philadelphia. I want to play a round of golf this afternoon. Please take me out to Merion (Country Club)." He would, at great inconvenience, have to take the dealer to lunch and out to play golf. But he didn't complain.

The issues of the 1960s were the same issues that had confronted the association for the past fifty years: fluctuating membership that caused cash flow problems, a change in location of the association's offices, a beginning discussion on the purchase of a permanent office site; expansion of the geographical area served by the association, and the concomitant discussion on an appropriate name change to adequately reflect the service area of the association; relations with the national association; an orderly transition in executive leadership; and, finally, expansion of services to member dealers in the area of education and insurance programs. We will look at each one of these issues.

The early 1960s was a period of economic stagnation, and the association's membership figures and budget projections showed it. The 1962 budget projected 320 members but the association only had 253. Given that type of membership decline, the projected budget of $102,419 would have a shortfall of over $21,000 and the Board of Directors began, once again, an aggressive membership drive. As part of that effort, they created the position of director of dealer relations and appointed Charles Graff, longtime MALA employee and assistant secretary/treasurer, to the position. Part of his compensation package was a 20 percent commission on any net gain in new members. These efforts had only mixed results, however, because by 1970 the association had only 291 members.

Part of the problem with membership was the dues structure of the organization. Throughout the decade, a variety of mechanisms were used. Early in the 1960s an effort was made to establish categories of membership that provided for increased services for higher dues category. This proved confusing to the dealers and a headache for the office staff to administer; thus it was abandoned. The organization would return, in 1967, to a sliding scale of dues with a minimum payment of $150 for dealers with sales volume up to $50,000 and increasing to $300 for those dealers with sales volume of $1,500,000. There would still be complaints, though, that the dues were too high.

Beginning in 1963, the Board of Directors periodically discussed relocating the offices of the association as a way of cutting costs. In particular, for the first time they began to discuss having the association purchase its own building to house the MALA offices and other tenants as a way of cutting expenses. The purchase of a building would not come for another decade but, in 1966, MALA moved from 2 Penn Center to the First Pennsylvania Bank building in Ardmore as a way of cutting rental costs.

At the same time, they looked to geographical expansion as one way of easing some of their problems. In particular, they actively pursued affiliation agreements with the Philadelphia Lumbermen's Exchange, the Western Pennsylvania, Northern New Jersey, West Virginia, and Virginia associations. Several of these efforts would prove successful in future years, although not immediately. However, one of these overtures, in particular, did not work out. The Northern New Jersey Association began to aggressively recruit members in Southern New Jersey and this caused a strain with the national association over recognition of geographic territories for each of the associations.

A picture of the January 1, 1948 meeting of the Retail Lumber Dealers Association of Maryland, a group that would eventually merge with MALA. Photo courtesy of Maryland Lumber Company

Interior of the Lehigh Lumber Company, circa 1950. The owner, Charles Krimm, is the man in the suit. Photo courtesy of Lehigh Lumber Company

First class of the Annual Retail Lumber Dealers Short Course, 1947, State College, Pennsylvania.
Photo courtesy of EBMDA

One of the ways that MALA was able to bring these other groups eventually into their association was through their participation in the various insurance programs offered by MALA. Having once gotten them to participate in one or two of the programs offered (usually an insurance program) by the association, it was very easy then to have them affiliate with MALA. In an effort to help this process, discussions were also held within the Board of Directors on name changes. Among those suggested in the early 1960s was the inclusion of "Building Materials Dealers" within the title, but that was rejected in 1961 as being "too confusing." By 1969, the "Name Committee" of the Board of Directors was suggesting "Atlantic Building Materials Association". However, that was never really discussed by the board.

The 1960s also began with MALA informing the National Retail Lumber Dealers Association (NRLDA) that they were resigning from the association. It was MALA's contention that the national association should be addressing only national and legislative matters and that NRLDA's current activities went far beyond those envisioned when it was created. In particular, MALA felt that many of the activities were impinging on their operations, that the structure of NRLDA was "cumbersome and outmoded." By 1961, after considerable correspondence, a meeting was set up between the MALA Board of Directors and the officers of NRLDA. The officers of NRLDA indicated that they thought most of the problems were ones of communication and MALA agreed to remain in the national association while efforts were made to address their concerns. However, in November 1962, the Executive Committee voted to resign from NRLDA after the national association took actions that had never been discussed by the constituent groups by announcing training sessions that had never been cleared before. Perhaps the most damning action, however, was NRLDA's proposing a building exposition in Atlantic City one month before MALA's convention in the same city. When confronted with this, the NRLDA officials indicated that they thought Atlantic City was in North Jersey. It also turned out that NRLDA had notified all of the regional associations except MALA of these actions. MALA would rejoin the national association, now renamed the National Lumber and Building Material Dealers Association (NLBMDA), in 1966 after several of its conditions were met. Relations would remain strained until 1967 when MALA member William E. Norman was elected treasurer of the national association. At this point debate over whether to stay in the national organization or not ended.

Top Management Workshop, 1953. This type of educational program was an integral part of the services provided by MALA to its members. A similar program continues today. Photo courtesy of EBMDA

As the decade of the sixties progressed, educational programs and a variety of insurance programs for the dealer members increasingly came to the fore as one of the main benefits of members. Educational programs were wide ranging and included one-day sales training meetings, two-day clinics on home improvements, weekend executive management workshops, estimating clinics, profit management workshops, store planning and design workshops, merchandising clinics, estate planning. However, by the end of the decade, dealers began to complain that the association's educational programs were not keeping up with modern times and the Board of Directors agreed to undertake an extensive "Opinionaire" of members in an effort to update the educational program of the association.

Insurance programs are dealt with in detail elsewhere. Suffice it to say for now that by the end of the decade it was reported that MALA operated the largest group insurance program of all of the federated groups within the national association. By this time, profit sharing, retirement, workers compensation, life insurance, and medical insurance were being offered through the association.

Finally, in discussing the decade of the 1960s, it is interesting to note

RETAIL LUMBER INSTITUTE

Jan. 18 - Feb. 12, 1954

OGONTZ CENTER, PA.
(*near Philadelphia*)

8th Annual
TRAINING PROGRAM FOR
THE LIGHT CONSTRUCTION
INDUSTRY

Sponsored by
MIDDLE ATLANTIC LUMBERMENS ASSN.
and the
EXTENSION SERVICES OF THE
PENNSYLVANIA STATE COLLEGE

GIVE YOUR EMPLOYES A FIRM HOLD ON THE FUTURE WITH YOUR INDUSTRY'S TRAINING PROGRAM

★

YOUR BUSINESS WILL PROFIT FROM THEIR SUCCESS

Cover of a flyer for the Retail Lumber Institute, January 1954. Another example of the educational programs offered by MALA.
Photo courtesy of EBMDA

that the association decided to celebrate its seventy-fifth anniversary by recognizing those members whose business had been in existence for seventy-five years or more and to have a post-convention anniversary trip to the Grand Bahamas on February 19–26, 1967.

The change in decades from the sixties to the seventies saw major changes in the association. Probably most important was the fact that two longtime senior staff members, Bob Jones and Charles Graff, either retired or were about to retire and replacements had to be found. Secretary/Treasurer Graff retired in 1969 and was replaced as secretary for a short time by longtime employee Betty Bunting. The treasurer's position once again became a dealer member. Russell J. Allen, a staff member who had been hired in the early 1960s to help "in the creative writing, research, bulletin development and all other promotional work necessary to the continued growth and service of the Association" and had served as assistant secretary/assistant treasurer since early in the decade, was elevated to the position of executive vice president. He held that office until his untimely death in January 1978. Taking over in 1971 as secretary was Harry H. Johnson III, a staff member who had been hired in 1969 as a field representative to work particularly with dealers in the New Jersey area. Harry continues with the association today as senior vice president.

The replacement of Graff and Jones was traumatic for the association because of their long tenure with MALA and because Jones had come to be known as the personification of MALA. He would continue, however, to conduct a number of educational programs for which he had become famous. These would be conducted throughout the country under the sponsorship of Owens Corning Fiberglas.

The difficulty for the association caused by these retirements is remembered by former MALA 1969 president and 1970 chairman of the board (the title was changed during his term of office) Joseph E. Haenn:

> It became time for Bob Jones to retire. While Bob agreed, I don't think he really wanted to let go. In retrospect, we were asking him to let go of the thing that had been the main and center ring of his life for many years. Not only that, there were financial considerations. The Association had never had a lot of money so part of Bob's compensation had been perks. He was given actual ownership of *Plan* magazine, and had been allowed various business expenses such as membership in the Union League. . . . The man had worked long hours and years for his dealers and for the entire industry. The Board had

recognized this, and had given the perks to Bob.

We, the Board, had a hard time trying to separate the part of the Association that was Bob's and the part that was not his. Everybody wanted to do what was right and fair but we were a bit lost about how to proceed. We were helped out of our dilemma by Owens Corning Fiberglas. They offered Bob sponsorship for various seminars across the country.

Bob Jones was named executive vice president emeritus and continued to attend some of the Board of Directors meetings and provide advice to the association. He also continued to conduct workshops and training sessions.

There was also another retirement around this time which affected the association. Frederick A. VanDenbergh, Esq., of the law firm of Saul, Ewing, Remick and Saul, and association counsel since 1945, retired. He was replaced by Lowell S. Thomas, Jr., who continues today as the association's counsel.

THE SEVENTIES AND EIGHTIES

One of the most important reorganizations in the association's history occurred in 1970. MALA, Inc., was created as a "management service organization of the Building Materials Dealers". The certificate of incorporation was dated September 21, 1970. MALA, Inc., was a corporation set up by the Middle Atlantic Lumbermens Association to carry out its business operation. It allowed the association to remain tax-exempt and non-profit. The officers of MALA, Inc., were the staff officers of the association, and the Board of Directors of the corporation were the Board of Directors of the association. In the certificate of incorporation, the purpose of the new entity was stated as: "To maintain and operate a trade information bureau for lumber and building materials dealers; to do each and everything necessary, suitable or proper for the accomplishment of this purpose; and to engage in any lawful act concerning any or all lawful business for which corporations may be incorporated under the Business Corporation law." In its registry papers, the new corporation indicated the kinds of businesses that it would conduct: "publication and

Photo courtesy of EBMDA

Dedication program, MALA headquarters office building, May 15, 1977. Photo courtesy of EBMDA

June 10, 1976

June 23, 1976

distribution of dealer directories, sale of forms, storefront construction and the management of conventions."

The Board of Directors of the Association/MALA, Inc, at an organizational meeting on September 29, 1970, elected Russell Allen, president; Harry H. Johnson III, secretary; and Edwin Scholtz, treasurer. The association board also passed a resolution having MALA, Inc., assume all personnel functions for the association, M.A.L.A. Group Insurance Trust, M.A.L.A. Retirement Trust, and the West Virginia Group Insurance Trust as of October 1, 1970. It also assumed the payroll of the association and the lease liability, published the *Dealers Directory and Buyers Guide*, and produced and staged the annual convention. "MALA, Inc., shall also manage the Business Forms Program, the Store Fixtures Program and future revenue producing programs that may be added with the consent of the Board of Directors." In other words, the new corporation took over all the obligations of the association as well as its revenue-producing operations and the services provided to members and they managed these programs for the benefit of the association. In return, they received a percentage of the dues paid. Initially, the association paid the corporation 90 percent of the dues income. Other entities associated with the organization, such as the Group Insurance Trust, paid 8 percent of premiums and the Retirement Program provided 50 percent of their net income. These amounts have varied over time and reflect the amount of work and services provided by the corporation for the various instrumentalities.

While there had been a previous corporate structure called Middle Atlantic Service Corporation, to provide certain services to the association, it had been primarily an operation to handle the writing of insurance polices. This corporate reorganization finally put the entire business end of the association and its various subsidiary operations on a sound financial basis and has proven to be a highly successful method of organizing the business, financial, and insurance operations of such an association. It also allows MALA, the association, to remain a nonprofit entity under the tax code.

During the early 1970s, "merger" talks were held with the Western Pennsylvania Lumber Dealers Association, the New Jersey Lumber and Building Material Dealers Association, the West Virginia and the Virginia associations. The New Jersey Association eventually backed out of the talks. The Western Pennsylvania Association then dissolved and their membership was taken over in 1972 by MALA, due in large part to the efforts of Bill Kildoo. Eventually, the assets of the Western Pennsylvania Association would be liquidated and more than $25,000 given to MALA to help in the construction of the new headquarters building for the association which is discussed below. The efforts to attract the Virginia and West Virginia associations would continue into the 1980s, but would prove unsuccessful.

At this point, with the new organizational structure in place and membership beginning to climb, the staff began to grow—by 1974 the association would have a paid staff of ten and was increasing yearly. By the time the association moved into a new building, the staff was handling services for almost 390 members. Serious efforts, therefore, had to be undertaken to look for a new building, owned by the association, to house all of its growing operations. A Building Committee was established in 1972 and the parameters set for its operation were: it was to be within a 5–10 mile radius of the current location; 3,000 sq. ft. to begin with, with the possible addition of 1,000 sq. ft.; and it was to be made out of wood to relate it to the industry. At the same time that the Building Committee was created, a fundraising drive was started. After looking in the Paoli/Malvern areas, the Building Committee reported that they had located a two-story building on Lancaster Avenue in Rosemont, Pennsylvania, that fit the conditions set by the Board of Directors. The association would occupy the basement and first floor of the building and the second floor would be rented to a tenant until needed. The board agreed to purchase the building. However, two days before settlement the seller rejected the sales agreement and the association began to look for a site all over again. Finally, at the June 20, 1975, meeting of the Board of Directors, it was agreed to purchase a parcel of land on Sloan Lane, just off of Baltimore Pike in Media, Pennsylvania, for $53,000. They also agreed to have the Building Committee select an architect and approve the design of the building. The building would be dedicated May 15, 1977, and the total cost of the building and furnishings amounted to $194,000 and involved no indebtedness. The entire cost came from dealer contributions and reserves held by the association for the new building.

With the dedication of the new headquarters building, a psychological barrier appears to have been passed for the association. The new building was a visible sign of the permanence of the organization and it is obvious from the dedication ceremony and the recorded reminiscences of past presidents that this was considered a major milestone for all those involved in the history of the association.

The excitement associated with the building and occupancy of the new headquarters building was tempered by the fact that the Justice Department was investigating MALA for compliance with the consent decree of 1941. The association had to supply the Department of Justice with copies of all publications and copies of the minutes of all meetings from

July 2, 1976

July 9, 1976

July 30, 1976

September 18, 1976

Dedicated May 15, 1977

1970 to 1975. It was also reported in April 1977, that the FBI had visited four or five dealers in relation to the probe. Finally, in August 1977, the Antitrust Division reported on the investigation:

> The investigation revealed that, on or about November 25, 1974, the Association mailed to its members a pamphlet entitled "Compensatory Pricing of Sheet Materials — Cut to Size," together with a transmittal letter urging the use of the pamphlet by members of the Association. This activity by the Association appears to violate that provision of the decree which prohibits the Association from:
>
> Formulating, promoting or participating in any plan or program to fix, determine, adhere to or to bring about the use of uniform markups, price differentials, allowances, discounts, or terms and conditions of sale with respect to lumber, lumber products or other building materials. . . .
>
> Information presently in our possession does not indicate that any dealer receiving the pamphlet made use thereof and, accordingly, further proceedings in this matter are not, in our judgment, warranted at this time. We will, however, continue to be alert to your Association's activities with respect to the consent decree.

Thus, the association escaped further legal entanglements. However, it was a lesson once again learned—MALA, like any other party to the consent decree and almost every other trade association in the country, had to be very careful concerning any action that might be construed to constitute restraint of trade.

Tragedy struck on January 23, 1978, when the executive vice president, Russell J. Allen, died suddenly. Allen had "grown up with the Association." He had first been hired as a photographer for the convention and *The Plan* magazine. Then he became secretary of the M.A.L.A. Group Insurance Trust and later was the developer and secretary of the M.A.L.A. Retirement Benefit Programs. At the time of his death he was the association's executive vice president and the first president of the wholly-owned subsidiary, MALA, Inc.

Following Allen's death, John D. Mitchell, who was the dealer-member chairman of the Board of the association, took on the added responsibilities of executive vice president of the association and president of MALA, Inc. A search committee was established to find a successor to Allen and by July 1, 1978, David B. Kreidler, was named executive vice president of the association and president of MALA, Inc.

Shortly after Kreidler's arrival, the association's insurance operation was reorganized with the creation of an insurance agency subsidiary, MARM, Inc., the Mid-Atlantic Risk Management, Inc., which allowed the new corporation "to operate a general insurance agency by acting as agents for insurance companies in soliciting and receiving applications for general life, health, accident, casualty, liability, and all other kinds of insurance." Also, the dues structure was revised and dues were raised to a minimum of $200 for sales volume under $500,000 and a sliding scale upwards to a maximum assessment of $375 for sales volume over $1,500,000.

The minutes of the Board of Directors of the association, in particular, as well as the minutes of the boards of the subsidiary corporations, become increasingly less fruitful sources of information throughout the 1970s and into the 1980s. As a result, it becomes increasingly more difficult to adequately deal with the history of the association. This is not because less is done, but rather more is being done by an increasingly larger paid staff, who make oral presentations to the boards which are not completely reported in the minutes. Also, it becomes apparent that the various boards are effectively using committees to transact business and make recommendations. As a result, the minutes of the boards contain much less detail, making the work of the historian that much more difficult and this history much less detailed. What would appear on the surface to be simply a question of less record keeping illustrates what was an important and fundamental change in the history of the association. Throughout the 1970s and 1980s, it was dramatically changing from a volunteer to a staff driven operation. While the various boards and committees provided guidance, it was the paid staff that increasingly ran the organization and provided the continuity of operation. This was vividly illustrated with the untimely death of Russ Allen in 1978. While the dealer-member chairman of the Board, John D. Mitchell, took over in name Allen's titles, it was the office staff that continued to run the day-to-day operation.

This has meant that the success of the association is increasingly dependent on a professionally trained staff. The current executive vice president, David B. Kreidler, and the senior vice president, Harry H. Johnson III, are both certified association executives. E. Thomas Fleck, the Vice President-Staff who oversees the Retirement Plan Trust is a certified employee benefits specialist and Richard W. Brown, Vice President-Staff, who serves as convention manager and director of education holds a certificate from the American Society of Association Executives, Conventions and Expositions Section.

A sense of the activities of the association can be obtained by the 1980

Edwin F. Scholtz, on left, Building Committee chairman greets Edgar B. Harman, chairman, EBMDA at Dedication Ceremony. In background is William Grover, an EBMDA employee.

With Edwin F. Scholtz, Building Committee chairman, and William Glover, EBMDA employee, holding a piece of lumber, Edgar B. Harman, chairman of EBMDA, officially opens the building by "sawing the board," rather than cutting the ribbon. Photo courtesy of EBMDA

OPERATIONAL CHART

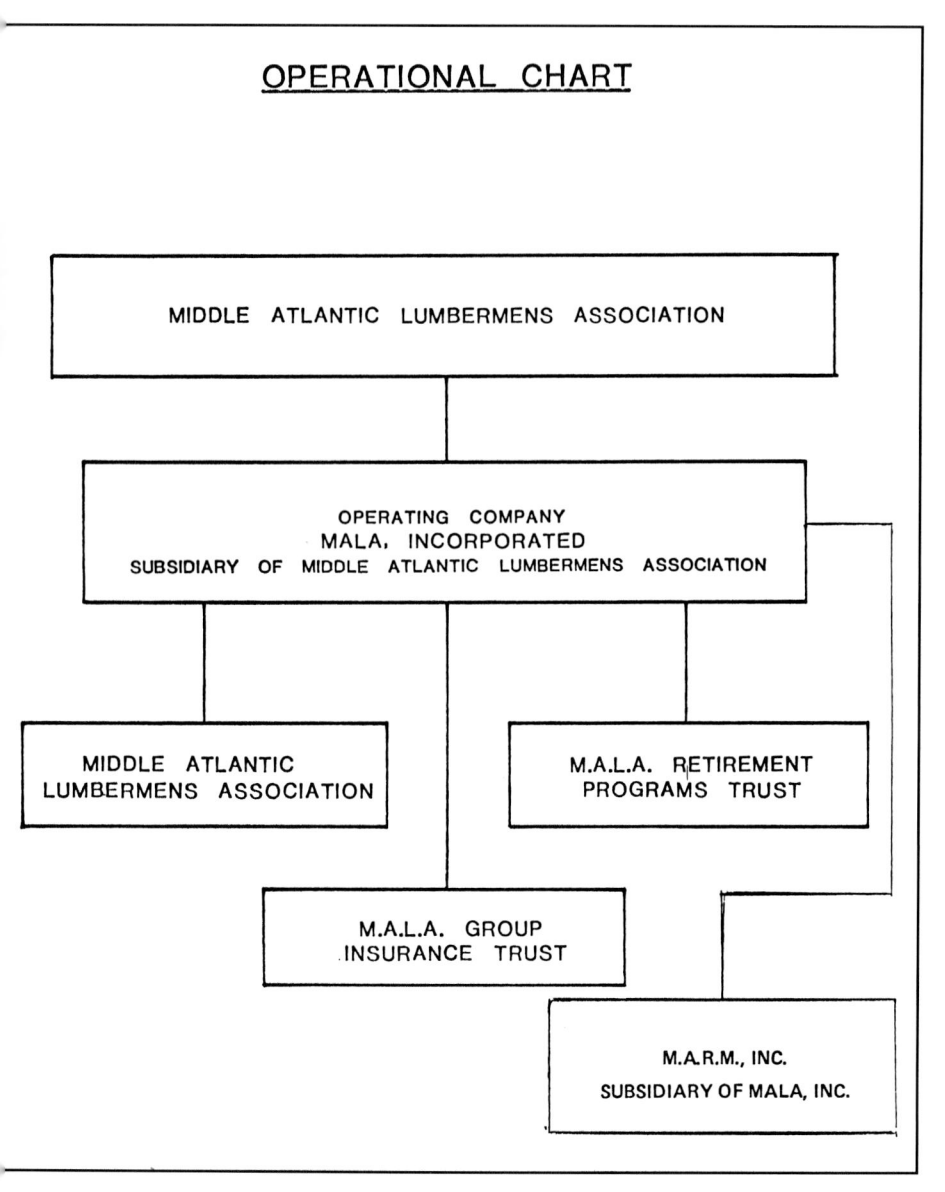

```
┌─────────────────────────────────────────────┐
│    MIDDLE ATLANTIC LUMBERMENS ASSOCIATION   │
└─────────────────────────────────────────────┘
                      │
┌─────────────────────────────────────────────┐
│            OPERATING COMPANY                │
│            MALA, INCORPORATED               │
│  SUBSIDIARY OF MIDDLE ATLANTIC LUMBERMENS   │
│                ASSOCIATION                  │
└─────────────────────────────────────────────┘
         │              │              │
┌──────────────────┐  ┌──────────────────┐
│ MIDDLE ATLANTIC  │  │ M.A.L.A. RETIREMENT│
│ LUMBERMENS       │  │ PROGRAMS TRUST    │
│ ASSOCIATION      │  │                   │
└──────────────────┘  └──────────────────┘
         ┌──────────────────┐
         │  M.A.L.A. GROUP  │
         │  INSURANCE TRUST │
         └──────────────────┘
                  ┌──────────────────┐
                  │   M.A.R.M., INC. │
                  │ SUBSIDIARY OF    │
                  │   MALA, INC.     │
                  └──────────────────┘
```

1983 organization chart of MALA

report of the chairman of the board, John D. Mitchell: Twelve seminars were held consisting of twenty-one days of instruction and 540 attendees. Nine different surveys were conducted and the legislative committee took positions on six different issues. There were 207 different locations covered by the insurance trust for a total of 3,175 employees.

By 1983, it became obvious that the name of the association should once again be changed to reflect the changes that were occurring in its operation. Thus a new set of Bylaws and a proposal to change the name was submitted to the membership. At the annual meeting in February 1984, MALA became the Eastern Building Material Dealers Association. At the same time, the membership also approved a Code of Ethical Standards.

Three other developments in the mid 1980s deserve mention. First, in 1984, EBMDA became associated with the *Building Material Retailer,* a publication sent to retail dealers throughout the country. To some extent, this publication filled a gap caused by the suspension of the publication *The Plan* and allows dealers to obtain information pertinent to the building material dealer. A special section highlights the activities of EBMDA.

A second development was the creation of the Education Foundation. The Board of Directors approved the creation of the Eastern Building Material Dealers Education Foundation on September 12, 1986. The reasons given for the creation of this foundation were fivefold:

    To improve the financial expertise of the independent dealer.

    To improve the marketing strategy and merchandising skills of the independent dealer.

    To encourage the short and long-term planning for the independent dealer.

To improve productivity through organization structure and systems analysis.

To improve efficiency through computerization.

Since its formation, the foundation has been actively soliciting funds and has raised a total of $118,605 to date. The foundation has been able to provide subsidies to the Management Development Program, the Lumber Short Course, and the Top Management Workshop.

The third development of note in the 1980s was the increasing reliance on strategic planning as a tool to be used in the operation and development of new programs for EBMDA. This effort was begun in 1984 and had a two-prong program: Help the dealer become a better businessman and to strengthen the association. To address the first part of the program, the Strategic Planning Committee recommended the creation of the Education Foundation mentioned above. To address the second part of the program, the committee submitted a report to the Board of Directors and it was approved at their September 14, 1990 meeting. This report called for improving existing EBMDA services, the growth of the association and recommended needed administrative changes. At the same time, the committee reported on the further encroachment of government into business activities, the need to expand EBMDA's influence in the governmental and regulatory arenas at the state and federal levels; examined the escalating health-care costs and the Eastern Group Trust as well as the Mid-Atlantic Risk Management and the Property/Casualty Programs and the annual convention, the Eastern Market.

While the activities of the Eastern Building Material Dealers Association, as it approaches its centennial, are covered in the subsequent chapter, it is fitting to end this history of the association in the 1980s with two examples. One illustrates the old axiom that the more things change, the more they stay the same: In the first decade of the twentieth century, the association went on record opposed to the conversion to the metric system. In 1984, the association went on record opposed to the conver-

EASTERN
Building Material Dealers Association

CODE OF ETHICAL STANDARDS

WE, the members of Eastern Building Material Dealers Association, pledge ourselves to faithfully serve our communities and our customers.

WE realize that for any industry to keep abreast of the ever changing times, it should continually strive to work in the interest of a general public.

WE pledge ourselves to develop and encourage high standards of integrity and service among those persons engaged in the lumber and building material business.

WE pledge ourselves to greater knowledge through education and modern marketing practices related to and helpful to this industry and our community.

WE pledge ourselves to broaden public understanding of the importance in our economy of the lumber and building material industry.

WE pledge ourselves to create, develop and maintain sound management principles, improved business methods and advanced operating techniques.

WE pledge ourselves to operate in a legal and ethical manner.

WE pledge ourselves to meet the reasonable needs of our customers and the protection of the consumer by full disclosure of the terms of the sale and finally,

WE pledge ourselves to promote a better understanding of the American, competitive, free enterprise system.

sion to the metric system. Of the two hundred delegates to the National Metric Council voting on the proposal, only three voted against converting to metric and one of the three was the EBMDA member, B. Harold Smick, Jr. The second example shows that change does, in fact, occur. In 1986, the first woman, Mary K. Rearick, was elected to the Board of Directors of the association. She represents one of the centennial members of the association, J. H. Rearick and Son, Inc., whose predecessor company, Atticks and Britcher, helped organize the association in 1892.

Meetings with governmental officials are another service provided by the association in an effort to keep dealer-members up-to-date on governmental matters. Left to right, Richard A. Kauffman, John H. Auld, Congressman William Goodling, Mary K. Rearick, and John H. Rearick discuss matters at the 1975 "Conference with Congress." Photo courtesy of John H. Rearick and Son, Inc.

National Lumber and Building Material Dealers Association
Publications from NLBMDA

FORMS

☐ **Standard NLBMDA Loss & Damage Claim Form**
Provides a record of all facts concerning the lost or damaged shipment. Pads of 50. **$1.50 mbr./$2 non-mbr.**

☐ **Product Complaint Form**
Record all necessary defective product information. Pads of 50. **$1.50 mbr./$2 non-mbr.**

TRAINING MANUALS

☐ **The Black Hole of Delivery**
How to conduct an in-house survey of delivery costs and services. Specify sales for appropriate volume: under $2 million and $2 - $5 million. **$45 mbr./non-mbr.**

☐ **Bottom Line**
A self-study product knowledge training program to increase sales, motivate employees, and maintain customers. Produced by the Northeastern Retail Lumber Dealers Foundation. Contains: *two newly updated manuals* (with chapters on lumber, millwork, doors/windows, plywood, oriented strandboard and waferboard, insulation, interior paneling, roofing, flooring, ceilings, siding, gypsum, paint, masonry cement, carpeting, and specialty products); and *administrators guide*; and *test booklet*. **$99 mbr./$150 non-mbr.** Test booklets only (with student instructions and administrators guide): **$39.95 mbr./ $60 non-mbr.** Note: you must have the recently updated 2 volume Product Manual on hand for students' use in completing the test.

☐ **Building Good Employee/Labor Relations**
How to manage with or without unions, comply with the Civil Rights Act, hire and fire employees, and more. **$10 mbr./$19 non-mbr.**

☐ **COIN**
Easy-to-administer questionnaire used to test potential cashiers. Set includes 5 tests, one key. **$15.75 mbr./$25.75 non-mbr.**

☐ **NLBMDA Computer Manual**
Contains a listing of vendors providing computer systems for building material retailers, a computer glossary, vendor profiles and much more. Revised and updated in 1988. **$29 mbr./$39 non-mbr.**

☐ **Construction Methods, Blueprint Reading, Material Takeoff**
15-book course teaching residential construction methods. **$175 mbr./$209 non-mbr.**

☐ **Designs for Success**
Principles and procedures for the successful layout and design of your lumber yard or retail store. **$150 mbr./non-mbr.**

☐ **Employee Incentive Plan Manual**
Contains actual incentive plan descriptions, in-depth discussion of plans, and step-by-step guidelines for their establishment. **$49 mbr./$69 non-mbr.**

To order any of these publications, check the box of the product(s) you desire. You will be billed for your order, plus shipping and handling. Send this page to the address listed below, or call **1-800-634-8645.**
NLBMDA - Publications / 40 Ivy Street, SE / Washington, DC 20003

NAME _____
COMPANY _____
ADDRESS _____
CITY/STATE/ZIP _____
TELEPHONE _____

☐ **NLBMDA Federal Regulation Compliance Manual**
An important resource for any business; contains information on COBRA Continuation Coverage, the Minimum Wage, and Federal Employee Laws. **$125 mbr./$175 non-mbr.**

☐ **Forklift Maintenance for Profit**
Instructions and sample forms to aid in proper maintenance and care of lift trucks. Track and improve your maintenance expenditures. **$19 mbr./$29 non-mbr.**

☐ **Guide To Developing Company Procedures and Personnel Policy Manual**
Contains hints on everything from job descriptions to salary administration to insurance plans. **$69 mbr./$99 non-mbr.**

☐ **CLBMDA Lumber and Reference Manual**
This pocket-size reference manual is a must for every employee. **Quantity 1-50 at $10 mbr./$15 non-mbr. per piece; over 50 at $8 mbr./$15 non-mbr. per piece;** optional correspondence course available at $2 per piece.

☐ **Product Training Manual**
This 2-volume manual offers training in product knowledge, and answers customer and technical questions. **$90 mbr./$150 non-mbr.**

☐ **What It Takes To Succeed In Sales**
Packed with real case histories and special industry studies, this unique guide shows you how to select and retain highly productive salespeople. **$22.95 mbr./non-mbr.**

VIDEOTAPES

☐ **Building Business I**
Customer service video brought to you by the Northeastern Retail Lumber Dealers Foundation. Includes 5 student handbooks, one administrator's guide, and one video cassette. **$49.95 mbr./$150 non-mbr.**

☐ **"The Forklift and You" Forklift Operator Training Program**
Consists of a twenty-minute training video cassette, 5 Operator Workbooks, 5 Operator ID cards, 5 NLBMDA Official Certificates of Completion, and one Leader's Guide. **$99 mbr./$149 non-mbr.** Additional books, ID cards and certificates are available.

☐ **New Employee Orientation**
A twenty-minute video cassette training program for new employees. Includes 5 Student Workbooks, and one Leader's Guide. **$109 mbr./$159 non-mbr.** Additional books available.

☐ **OSHA Hazard Communications Program Compliance Manual**
Contains step-by-step instructions and all necessary materials for compliance with the HazCom Rule. Includes 20 minute video training cassette. **$100 mbr./$175 non-mbr.;** state regulations where applicable are $25 additional.

☐ **"A Profile of The Professional Remodeler"**
15 minute video perfect for suppliers, lumber dealers, and homeowners. **$29.95 mbr./$39.95 non-mbr.**

☐ **Selling Skills and Customer Relations**
Two hours of video instruction in basic selling skills and customer relations. Contains nine lessons on three video cassettes, 5 Student Workbooks, and one Leader's Guide. **$395 mbr./$595 non-mbr.** Additional books available.

☐ **"Your First Day" Employee Safety Training Video**
Designed for the orientation and training of your new employees. Keep safety a part of your yard and store policy! **$29.95 mbr./ $39.95 non-mbr.**

Magazine ad published in Building Material Retailer, *April 1991.*

MALA headquarters office building

Chapter IV
Eastern Building Material Dealers Association Today

In the fifty years of operations in the name of Middle Atlantic Lumbermens Association, the governance of the association and its entities transcended the typical voluntary association. In its formative years, many activities were directed and operated by volunteers. As the budget and staff grew, as a result of a growing membership, additional services were added that would be of even more benefit to members. For example, in 1970 a professional staff of seven persons was almost evenly split between Group Insurance operations and the association. By 1990, a professional staff of twenty-two had become responsible for the day-to-day operations on behalf of the Group Insurance Trust, the Retirement Programs Trust, a Member Services department and produced over $700,000 per year in sales of business forms, merchant agreements on credit cards and publications. In addition, new ventures included the installation of a DOT Drug Testing Consortium and four Political Action Committees at the state level and an Educational Foundation.

The measurement of progress for this organization is not limited to a mere counting of its members. Today, EBMDA services have become critically important to the industry. Many members have expressed that in lean years, many times the dividend from association programs meant the difference between a profit and a loss for their firm. However, the focus of

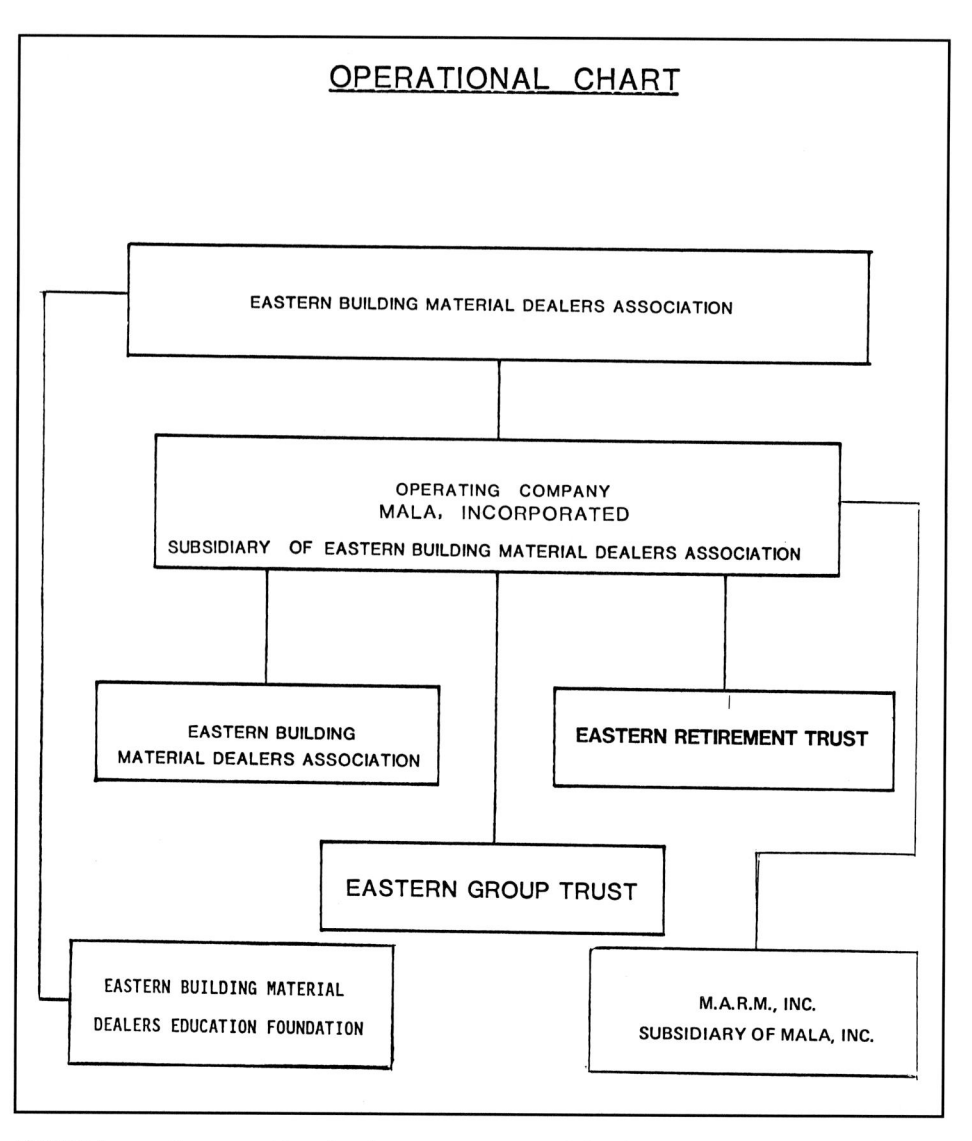

OPERATIONAL CHART

```
EASTERN BUILDING MATERIAL DEALERS ASSOCIATION
                    |
        OPERATING COMPANY
          MALA, INCORPORATED
SUBSIDIARY OF EASTERN BUILDING MATERIAL DEALERS ASSOCIATION
                    |
        ┌───────────┴───────────┐
EASTERN BUILDING          EASTERN RETIREMENT TRUST
MATERIAL DEALERS ASSOCIATION
                    |
           EASTERN GROUP TRUST

EASTERN BUILDING MATERIAL           M.A.R.M., INC.
DEALERS EDUCATION FOUNDATION        SUBSIDIARY OF MALA, INC.
```

EBMDA continues to be the improvement of the business position of building material dealers. Almost 150 persons, through the use of standing committees representing the membership, provide guidance for the association staff in its attempts to lead the activities of the industry.

1991 COMMITTEES

Nominating Committee: Chairman Aldo Braido, General Supply Company

Bylaws Committee: Chairman B. Harold Smick, Jr., Smick Lumber & Building Materials

Legislative Committee: Pennsylvania Chairman Terry L. Kauffman, Reinholds Lumber & Millwork
New Jersey Chairman B. Harold Smick, Jr., Smick Lumber & Building Materials
Maryland Chairman Robert M. Bushey, Cavetown Planing Mill

Delaware Chairman J. Fred Robinson, Newark Lumber Company

Education Committee: Chairman John H. Eaton, Jr., Barrons Enterprises, Inc.

Business Development Committee: Chairman Gregory Shelly, Shelly & Fenstermacher

Membership Development Committee: Chairman O. Grant Little, Little Lumber Company

Management Development Committee: Chairman Randall Brunk, People's Supply Company

Convention Advisory Committee: Chairman Vincent J. Tague, Tague Lumber Company

Associate Member Committee: Chairman James P. Rauch, Crafton Lumber Company

Building Committee: Chairman Edwin F. Scholtz, Sykes-Scholtz-Collins

In addition to these standing committees, special committees, such as the Centennial Committee chaired by B. Harold Smick, Jr., of Smick Lumber & Building Materials Center, and Strategic Planning Committee, Chairman Bruce C. Ferretti, Lehigh Lumber Company, have been created for special needs.

Other related organizations sponsored or authorized by Eastern Building Material Dealers Association include the operating subsidiary MALA, Inc., Chairman Vincent J. Tague; its subsidiary Mid-Atlantic Risk Management, Inc., Chairman David Waitz; Eastern Group Trust, Chairman Jay F. Risser; Eastern Retirement Trust, Chairman Perry E. Brunk; and Eastern Building Material Dealers Education Foundation, President Frank M. Hankins, Jr. The association itself is chaired by James P. Rauch, Crafton Lumber & Supply, for 1990 and 1991, and the vice chairman is Bruce C. Ferretti, Lehigh Lumber Company. Eighteen additional individuals, who serve rotating terms of two years, and the three senior Trustees from the Group Insurance and Retirement Trust, comprise the twenty-six voting Board Members.

Measured as the business world makes comparisons, EBMDA produces total annual revenue in excess of $20 million per year. In addition, the Property/Casualty Program revenues would add another $8.8 million but are not counted herein due to the direct deposit of such funds with the managing broker. By comparison, total revenue in 1980 represented $7.5 million.

Below are descriptions of some of our activities by entity as we enter our Centennial Year:

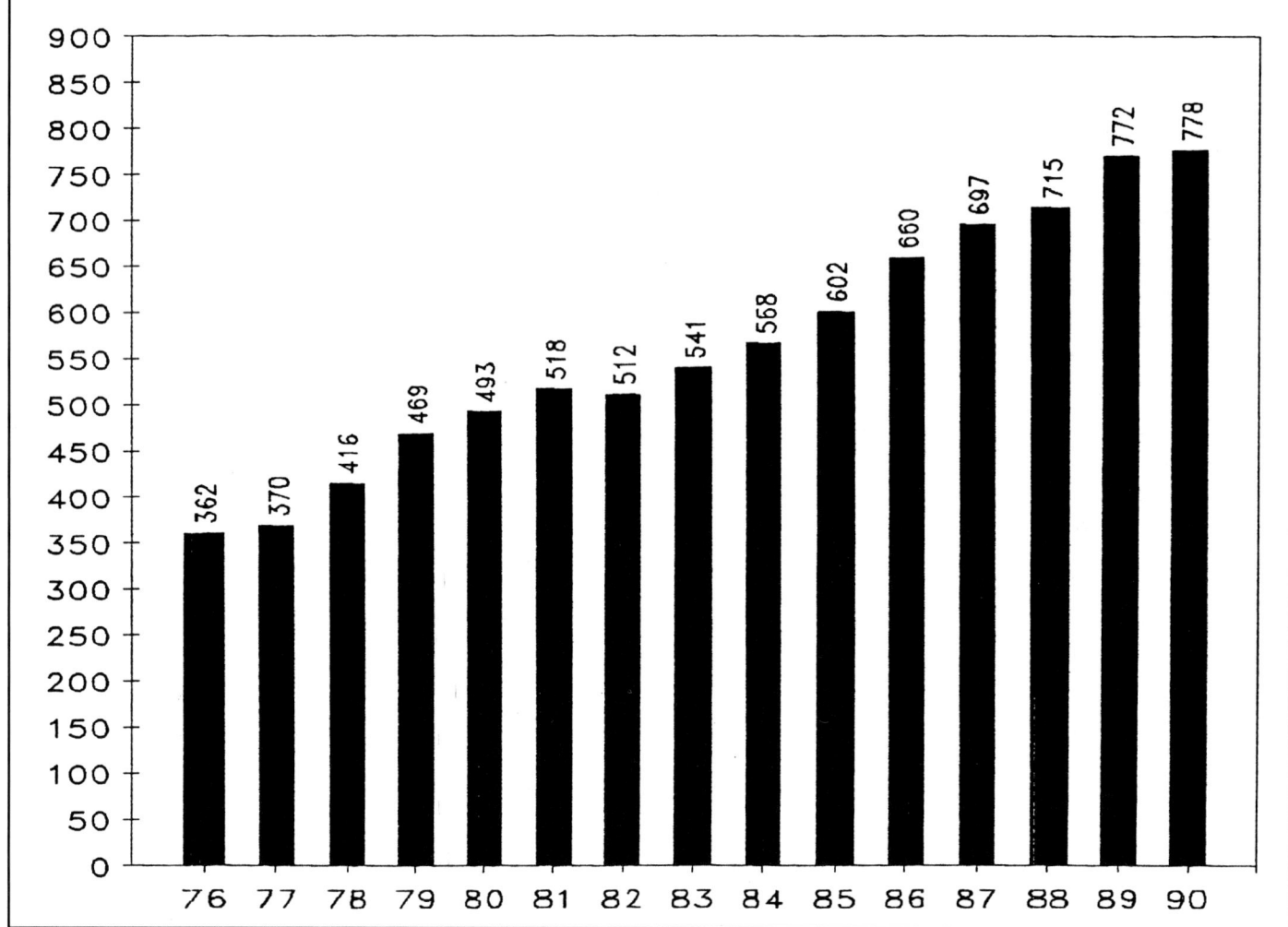

EASTERN BUILDING MATERIAL DEALERS ASSOCIATION TOTAL MEMBERSHIP

Year	Membership
76	362
77	370
78	416
79	469
80	493
81	518
82	512
83	541
84	568
85	602
86	660
87	697
88	715
89	772
90	778

EASTERN BUILDING MATERIAL DEALERS ASSOCIATION

The number of independently owned building material retailers continues to decline due to increased competition from publicly traded companies and a recessionary economy. In spite of that, retail membership closed 1990 with 549 active memberships, of which 96 were branches. Continued mergers and consolidations are expected to increase the number of branches of major members. There were 229 associate members reported for the close of business of 1990, yielding a total membership of 778 locations serviced by the association. By comparison, at the time of the name change to Eastern Building Material Dealers Association in 1984, 458 active members and 110 associate members participated.

In terms of dues, a total of $151,194 were paid in 1984 as compared with $235,863 in 1990. This increase was due not so much to any revision in the dues schedule, but rather to increased sales volume and to an increasing number of members participating in association activities. By the end of 1990 member equity had increased to $748,915 and to a cash position of $307,725. With an operating budget in excess of $2 million for the entire enterprise in 1991, it is evident that the association relies on dues to represent only a minor part of its total revenue picture. Much of the member equity results from a conservative appraisal of the value of the building, wholly owned by the association and rented by the service corporation subsidiary. It is interesting to note that the association employs no staff of its own. Rather, all staff activities are provided by MALA, Inc., on a service contract basis for a fixed percentage of dues revenue.

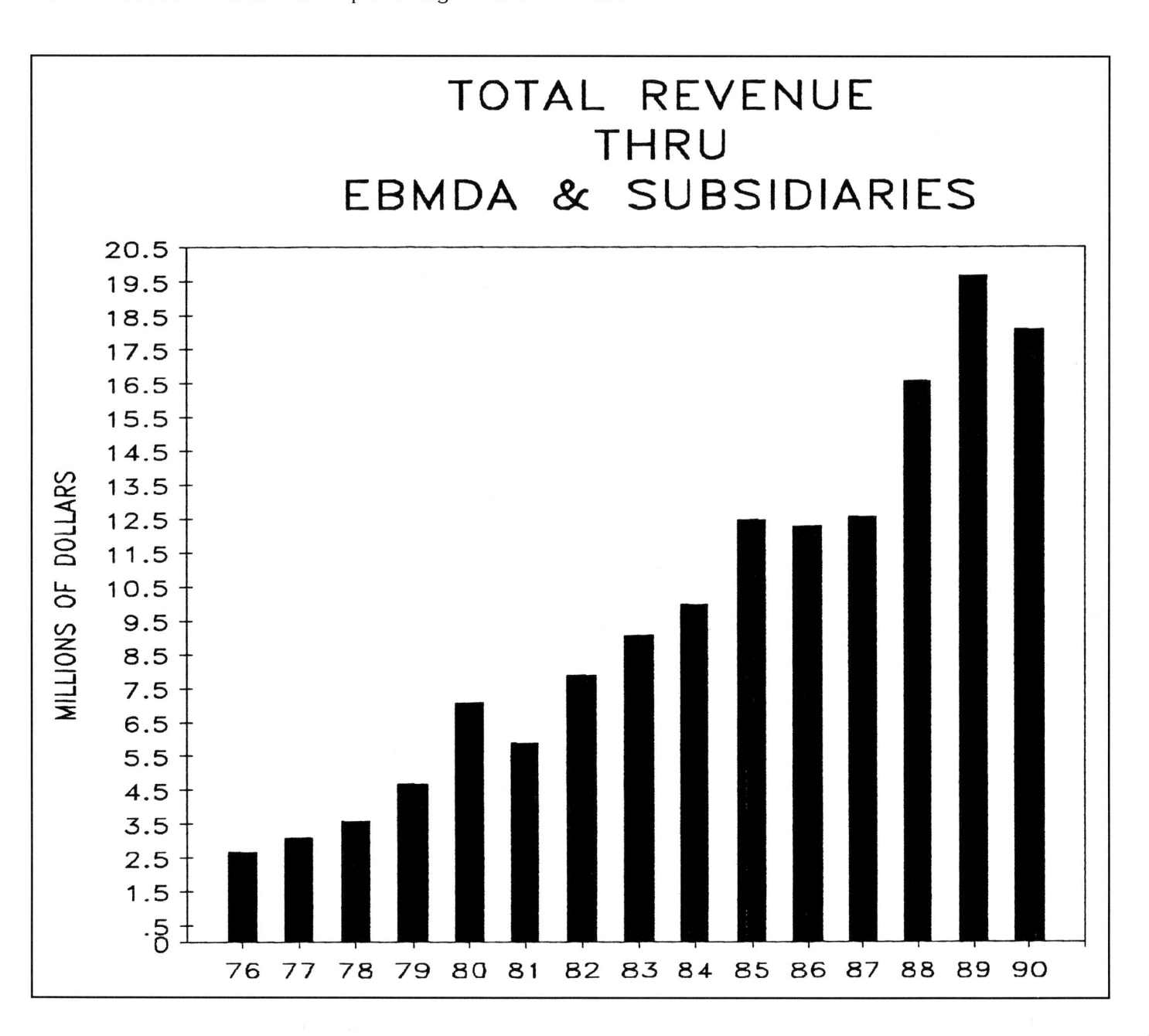

TOTAL REVENUE THRU EBMDA & SUBSIDIARIES

Specific activities, programs and services provided by the association after one hundred years also reveal the growth and progress that has been made.

First is legislative activity. *The Membership Bulletin* attempts to keep the membership informed of respective state legislation that can affect their businesses. Issued twenty-six times a year, it helps members identify issues of importance to them. Since there is a plethora of retail and business organizations that lobby state legislatures, the association has not participated in direct action in individual state houses. Most lobbying activity takes place in concert with other business leagues and coalitions with similar interests on major issues. However, in order to gain greater visibility at the state level, Eastern has sanctioned the organization of political action committees for the purpose of supporting certain candidates for state positions. The Political Action Committees for the respective states are:

- Maryland Lumber Dealers for Good Government, Inc.
- Pennsylvania Lumber Dealers for Good Government, Inc.
- Delaware Lumber Dealers for Good Government, Inc.
- Garden State Lumber Dealers for Good Politics, Inc.

By comparison with other major business organizations, the political action efforts of EBMDA are limited by the size of the balances available. Each state Legislative subcommittee is polled on an annual basis to obtain names of worthy candidates.

Education and training is another important activity of EBMDA. Continuing education is available through the Management Development Program and is designed for younger generation employees with management potential. This ongoing process meets twice a year

---

ROUTE TO:
General Manager
Store Manager
Sales Manager
Personnel
Accounting
Other

EASTERN Building Material Dealers Association
604 East Baltimore Pike • Media, PA 19063 • FAX-215-565-0968 • Telephone-215-565-6144
Editor: Harry H. Johnson III, CAE

AFFILIATED WITH
NLBMDA
National Lumber and Building Material Dealers Association

March 6, 1991

GROUP TRUST DIVIDEND

The Trustees of the Group Insurance Program managed by EBMDA, have declared a Dividend for Group Term Life, Accidental Death and Dismemberment and Weekly Disability Income. Participating Dealers received notice of the Dividend in the form of a credit against the March Premium Statement. The Dividend was calculated on the basis of 10% of the 1990 Annual Premium for each of the respective coverages. For further free information regarding the Eastern Group Trust Plans, contact Joe Bradley at 215-565-6144.

The Trustees of the Plan are very cognizant of the medical care inflation trend evident nationwide. They have retained a consultant to examine ways to attempt to reduce this trend in the Eastern Group Trust. For 1990, Eastern Group Trust increased the Medical Plan Rates an average of 15% as compared to 21.6% evident in recent Wyatt and A. Foster Higgins Surveys. The newly released U.S. Chamber of Commerce Report based upon 1989 Survey results, reported an average of $2,853 per employee for medically related insurance costs, 9.3% of payroll. Total benefit costs reported by the U.S. Chamber study now reached 37.6% of payroll for 1989, including pension, vacation, sick pay and employee benefit plans. The trend continues in 1990.

WAR'S OVER--ECONOMIC REBOUND COMING

Thank God that one of the shortest invasions in history kept the U.S. Press occupied for the last six weeks. Several remarkable things became evident:

* The President's approval rating hit a new high as a result of his tough stance regarding a world dictator and his "hands-off" approach for the Military effort.

[cont'd]

in two-and-one-half-day sessions with a self-directed agenda. The association has also sponsored an annual "institute" conducted at Penn State University for the purpose of acquainting those new to the industry with training skills and selling techniques. The total cost for a week of such concentrated training typically runs between $500 and $600 per student.

At a more advanced level, the association sponsors its Annual Management Conference. This is usually held in some resort setting and its purpose is to sharpen the management skills of executives of member firms.

In addition, throughout the year, a series of one-day seminars are held covering topics of interest. The topics include construction estimating, sales techniques, DOT compliance and other business issues. Attendance at these sessions is typically 25–30, but combined, they represent an annual training budget that grosses $65,000 or more. The chart on the opposite page plots the revenues generated by the Education and Training programs through 1990. Eastern is noted for its heavy involvement in education and training with one of the most aggressive schedules within the industry. A recent sign of the importance of education to the association is the creation of the Education Foundation, which has as one of its goals the expanding of educational opportunities by subsidizing various training programs.

ROUNDTABLE DISCUSSIONS HAVE ALWAYS BEEN A GREAT WAY OF SHARING IDEAS AND PROBLEMS, AND GETTING TO KNOW THE BUSINESS BETTER. STIMULATING TOPICS, FACILITATED BY MR. BILL LEE, WILL MAKE THIS AN ENLIGHTENING EVENING.

"High Impact Concepts to Grow Your Business and Increase Your Margins in the 1990's"
* Learn ways to avoid getting off to the wrong start.
* Understand productivity standards you should shoot for.
* Develop steps to avoid a commodity mentality.
* Avoid the big mistakes most contractor sales people make.
* Investigate new marketing concepts you need to know about.

"Results Driven Sales Management"
* Measure effective/ profitable sales performance.
* Learn six areas of "Rewarding" responsibilities for the sales manager.
* Understand the sales manager's "Team Building" responsibilities.
* Recognize the sales manager's "Training and Development" responsibilities.
* Develop and utilize a "Sales Activities" program.
* Learn the 10 x 10 Truth Chart.
* Comprehend horizontal and vertical account penetration.
* Establish key account management.
* Develop quality circles in sales.

"Credit Management Strategies that Work in a Weak Economy"
* Learn how to spot a bad risk.
* Develop reference checking.
* Discover what the contractor's financial statement can reveal.
* Gather collection techniques.
* Know what credit files should contain.
* Develop a fool proof follow-up system.
* Avoid the mistakes most credit managers make.
* Pinpoint credit responsibility.
* Protect credit policy and customer service.

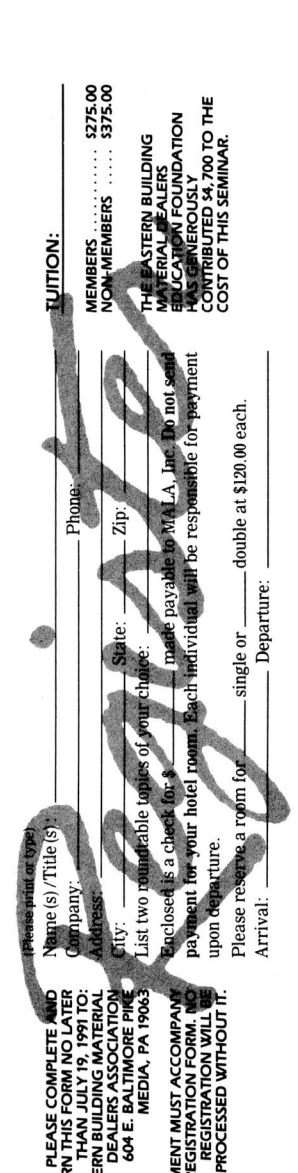

TUITION:
MEMBERS $275.00
NON-MEMBERS $375.00

THE EASTERN BUILDING MATERIAL DEALERS EDUCATION FOUNDATION HAS GENEROUSLY CONTRIBUTED $4,700 TO THE COST OF THIS SEMINAR.

EDUCATION & TRAINING REVENUE

Bar chart showing education and training revenue from fiscal years 75-76 through 89-90, with values ranging from 0 to 70000. Revenue shows an overall upward trend, peaking around 60000 in 88-89.

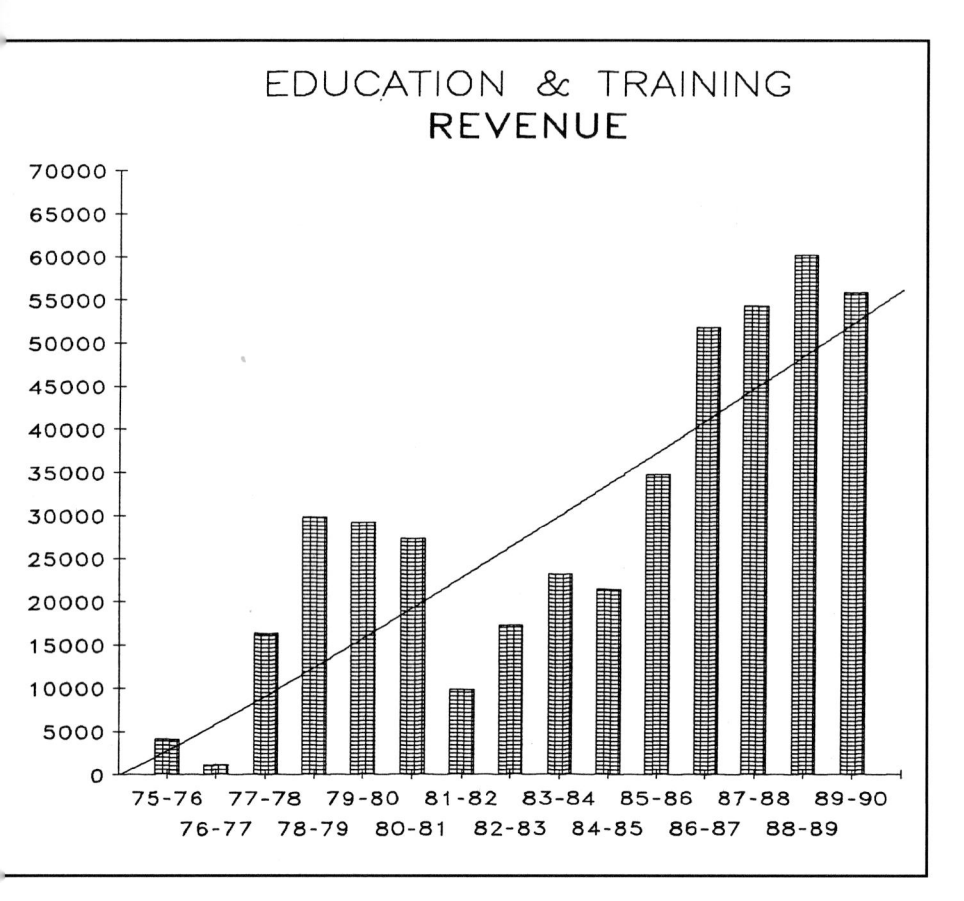

EASTERN GROUP TRUST

From humble beginnings in 1949, Eastern Group Trust has developed the reputation for consistently good service, fair dealing, and funding stability within the scope of medical care plans. Directed by a seven-member Board of Trustees, the Group Insurance Trust is managed using staff employees for the purpose of sales, installation, certificate and identification card issuance, billing and collection of premiums, payment of claims and providing reports for trustee and insurance carrier purposes. The administrator of the program is Harry H. Johnson III and the Group Insurance Department Supervisor is Joseph C. Bradley, Jr.

With the advent of computers using industry-standard third-party administration software, the staff has been able to process over 30,000 claims payments in 1990 and maintain the records on over 10,000 persons. Insurance functions for the Group Insurance Trust have been met since the inception by State Mutual Life Insurance Company, Worcester, Massachusetts, in which the Eastern Group Trust now represents the largest association client. The trustees control the funding for Eastern Group Trust under the guidance of industry consultant John B. Walton.

The 1991 budget approaches $11 million, notwithstanding the negative impact of staff reductions among the 270 participating employers. The trustees are elected by the participants pursuant to the requirements of Section 501(C)(9) of the Internal Revenue Code. The three most senior trustees become automatic voting members of the association Board of Directors, and the association chairman and vice chairman become ex-officio trustees to maintain continuity of communication between the association and the Board of Trustees.

Primarily organized to respond to the short-term disability income responsibilities under New Jersey statutes, the Group Insurance Trust has expanded its variety of coverages to include Group Term Life, Accidential Death and Dismemberment, Weekly Disability Income, Long-Term Disability Income, six medical plans and a dental plan. Continued participant growth is largely a result of face-to-face contact by association representatives working with prospective employers. Eastern Group Trust has been an innovator among association benefit plans in the institution of geographic-zone rates and modified age grading. Eastern Group Trust Plans do not contend that they are the "cheapest plans on the block" but have consistently attempted to make the premium ratings sensitive to differences in age profiles and geographic locale. The major objectives of the Group Insurance Trust have been to use plan designs that are easily understood by participating employees and to provide as much stability in funding as can be obtained in a rapidly inflating market place of medical care. The program is run as an "experience rated contract" with State Mutual, with surplus funding available for reallocation to reduce future premiums paid by employers and employees. The trustees exercise discretion in the distribution of these performance dividends, but have tended to allocate such funds on a pro-rata basis.

With the continued diversification of employer objectives, and with the increased establishment of state and federal government mandates in the employee benefits area, major hurdles remain for the continued growth of the Group Insurance Program. More and more voices are being raised in support of government-funded health programs in the future. As with other major services, the plan trustees, under the chairmanship of Jay F. Risser, have continued to revise plan design and operating methods to cope with this rapidly changing field.

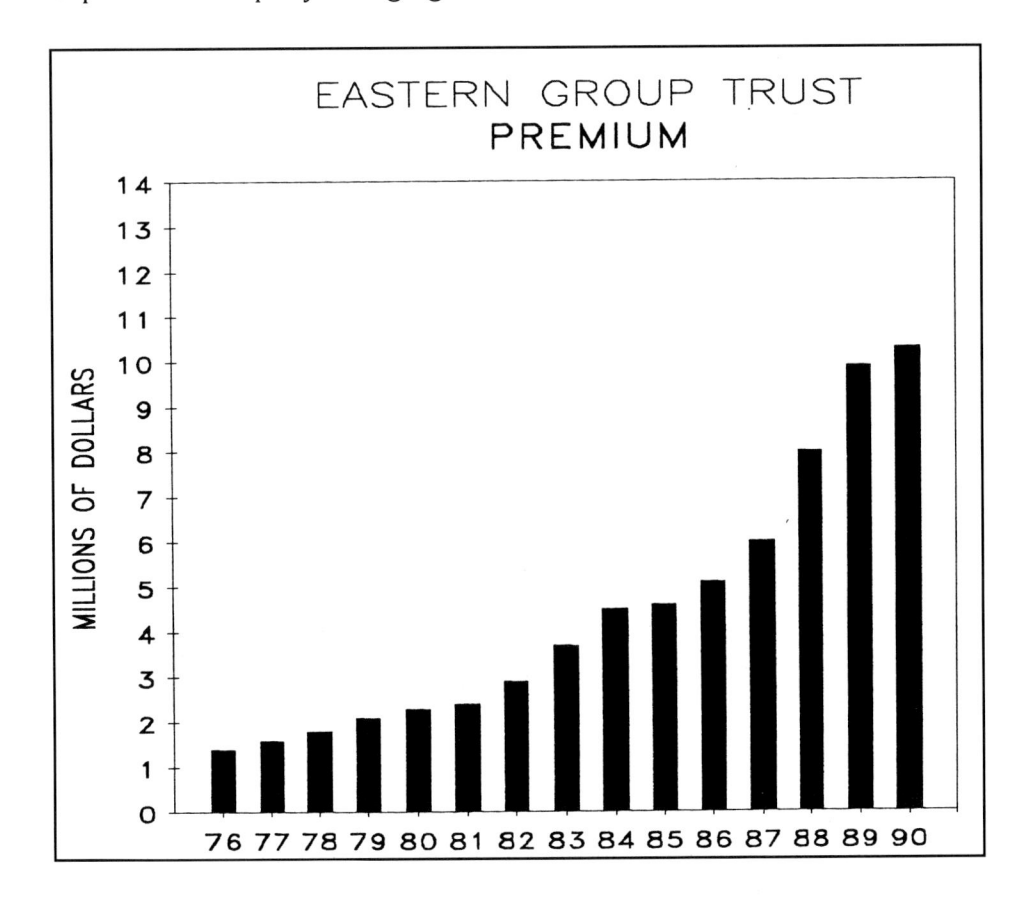

Group Benefit Plans

Life, Health, Disability, Dental, and Tax Advantaged Programs to Meet YOUR Special Needs Today; Estate Planning, Key Man, Buy-Sell, and Other Programs to assure YOUR Future.

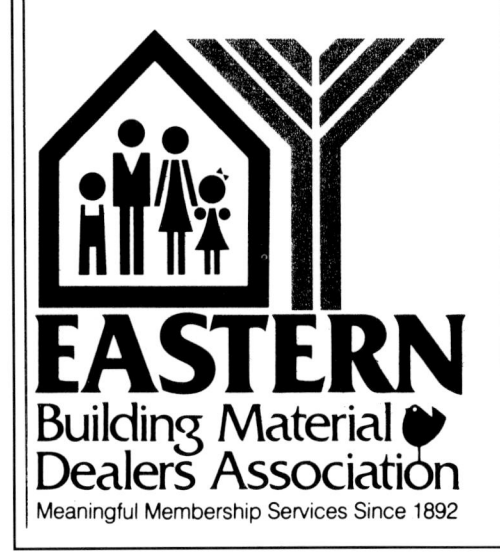

EASTERN Building Material Dealers Association
Meaningful Membership Services Since 1892

EASTERN GROUP TRUST PREMIUM

Year	Millions of Dollars
76	1.4
77	1.6
78	1.8
79	2.1
80	2.3
81	2.4
82	2.9
83	3.7
84	4.5
85	4.6
86	5.2
87	6.1
88	8.1
89	10.0
90	10.4

EASTERN RETIREMENT TRUST

Visionaries on the board may never have guessed the importance of the organization of the Retirement Program in 1954. Beginning with the Group Annunity Concept of retirement planning, a transition has occurred over the years which has caused the participation to increase from two employers to 88 in 1990. In 1964, the management of the retirement plans became "trusteed" by the Trust Department of First Pennsylvania Bank. In 1973, the Group Insurance Trustees, who served as the governing body for the retirement programs in the early years, saw the need to establish a free-standing trust with three major plan designs: Defined Benefit, Defined Contribution, and Profit Sharing Plans.

As with the Group Insurance Program, the Retirement Trust is supervised by a seven-member Board of Trustees elected by the membership, who serve in rotation. In addition, the chairman and vice chairman of the association serve as trustees ex-officio, and the three senior trustees serve as voting members of the association Board of Directors to foster interorganizational communication. Staff management of the Retirement Trust is in the hands of E. Thomas Fleck, administrator, and Harry H. Johnson III, assistant administrator. There is a staff of two administrative assistants who operate the computer software produced by Datair Management.

With the rapid growth of plan assets, the trustees have been able to expand the investment opportunity afforded participants. From one investment manager in 1970, the trustees have added additional professional fund managers to the current four who are now required to diversify the portfolio exceeding $36 million in assets. More then 3,000 employees are maintained in the computer records of the 108 employer plans.

The dual objectives of the Retirement Trust are to enhance the consistent investment return experienced by the plans by the pooling of assets, and to reduce the administrative costs encountered by small employers managing such plans. For example, virtually every major federal tax act had modified a requirement of retirement plans by the end of the decade of the 1980s. The trustees were able to consolidate all plan amendments into one legal project by retaining a legal firm and subdividing the cost among the participating employer plans. It is estimated that this reduced the legal fees required for compliance with these tax revisions to approximately one-half of what would have been normally paid by an individual employer.

The unique aspect of this EBMDA plan has not generally spread through the association community at large. In fact, contacts through the American Society of Association Executives indicate that the majority of associations that sponsor any type of retirement plan do so simply by endorsing an insurance plan or a professional trustee firm. With continued strong emphasis being placed on small business retirement programs at the federal level, it is likely that relaxed rules will be permitted in the management of such small business plans. Eastern Retirement Trust is positioned to capitalize on that

Retirement Trust is positioned to capitalize on that circumstance, should it occur.

Even with computerization, the sophistication of retirement plan management has caused the Trustee Board to consider the adoption of specialized sub-contractors to manage the 401(K), multiple option plans. While such challenges test the imagination and adaptability of the trustees and staff, Eastern Retirement Trust is positioned to be a strong industry player in the rapidly growing field of retirement plan management.

MID-ATLANTIC RISK MANAGEMENT

Through its history, the minutes of the association contain much evidence of the heavy impact that insurance costs have had on the members. The creation of the Pennsylvania Lumbermen's Mutual Insurance Company helped to reduce some of these costs, but not all. With the incorporation of MALA, Inc., in 1970, the board and staff continued to look at other opportunities for non-dues sources of revenue. Following a detailed search process, the association endorsed the Dodson Insurance Group for workers' compensation policies. Dodson committed itself to call on all the members of the association each year concerning participation in a plan. During this decade, no formal program was contemplated in the property insurance area due to the excellent rapport between PLM and the membership. Participation in the Workers Compensation Program managed by Dodson Insurance Group also stabilized at a hundred members.

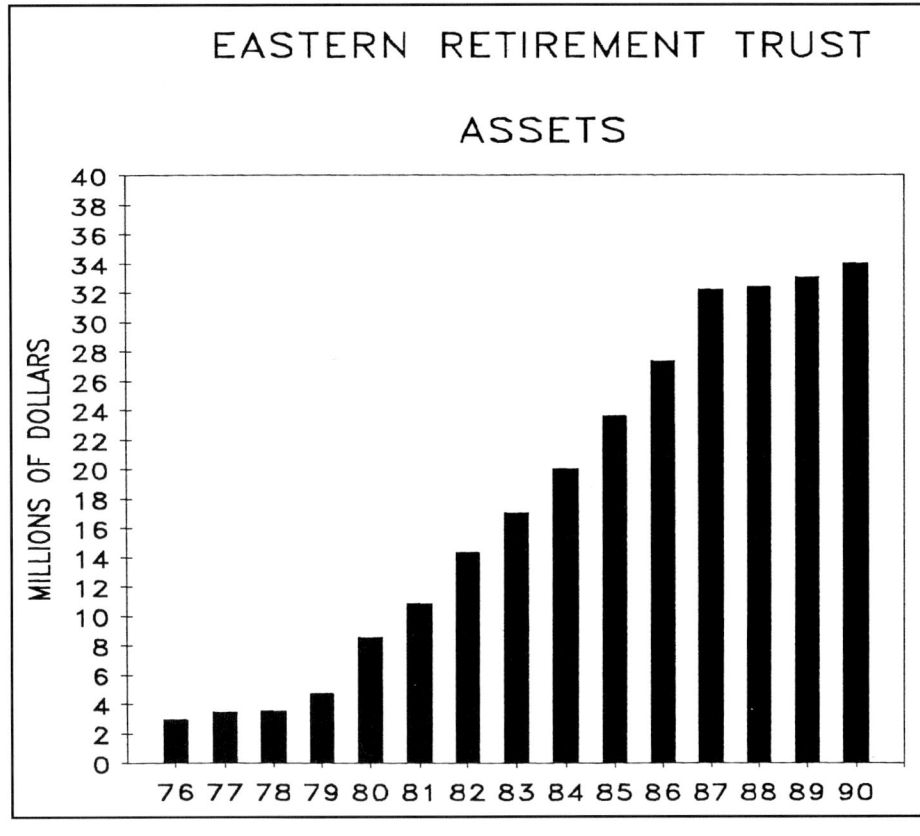

In 1978, it became evident that dealers were becoming uncomfortable with the standard fire insurance programs offered by Pennsylvania Lumbermens Mutual Insurance Company. A sub-committee of the MALA, Inc. board was established to work with Harry H. Johnson III to investigate and recommend revisions to the existing arrangements so that members could designate a local broker to receive commissions and experience rated dividends in the full line of property and casualty coverages. Following an extensive search process and after interviewing multiple insurance arrangements, including PLM, the subcommittee recommended the formation of a subsidiary corporation responsible for a newly designed Property/Casualty Program arranged through the Advanced Risk Management Division of Corroon & Black of Nashville, Tennessee. The marketing of this new program was to be handled by a Corroon & Black division in Washington, D.C., which made a commitment to call on all members and non-members of the association on behalf of the association program. However, they were to allow a participating member to select a local broker if he/she so desired.

During the next three years, the arrangements with Corroon & Black did not meet the expectations of the Mid-Atlantic Risk Management Committee. Because they failed to expand the program, the board authorized the investigation of alternative underwriters and managing brokers. The selection of Noyes Services in 1984 to perform the administrative functions provided the opportunity to use two underwriters in various aspects of the property casualty program. The PMA Group of insurance companies was chosen for the safety group in workers compensation coverage and commercial automobile coverage. Pennsylvania Lumbermen's Mutual Insurance Company was chosen for a special multi-peril program. The result of these changes was an immediate and dramatic upturn in participation.

During the six years that followed, participation more than doubled to 212 participating dealers and the number of insurance contracts expanded to more than 600. Annual premiums in the program now exceeds $8.8 million, roughly split between the two underwriters. Due to the skills of the underwriter, and the limitation of participation to employers who seek to have a better-than-average approach to safety, the success of the program has manifested itself in the form of substantial dividends—almost $1.5 million in dividends will be distributed to the participants for the year 1991.

In addition to the Property/Casualty Program, Mid-Atlantic Risk Management now is expanding its services into the creation of DOT Drug Testing Consortium for compliance with federal DOT regulations for truck drivers. In addition, additional policies are negotiated on behalf of members for integration with flexible benefit plan, 401(K) plan options and variable payroll deduction plans.

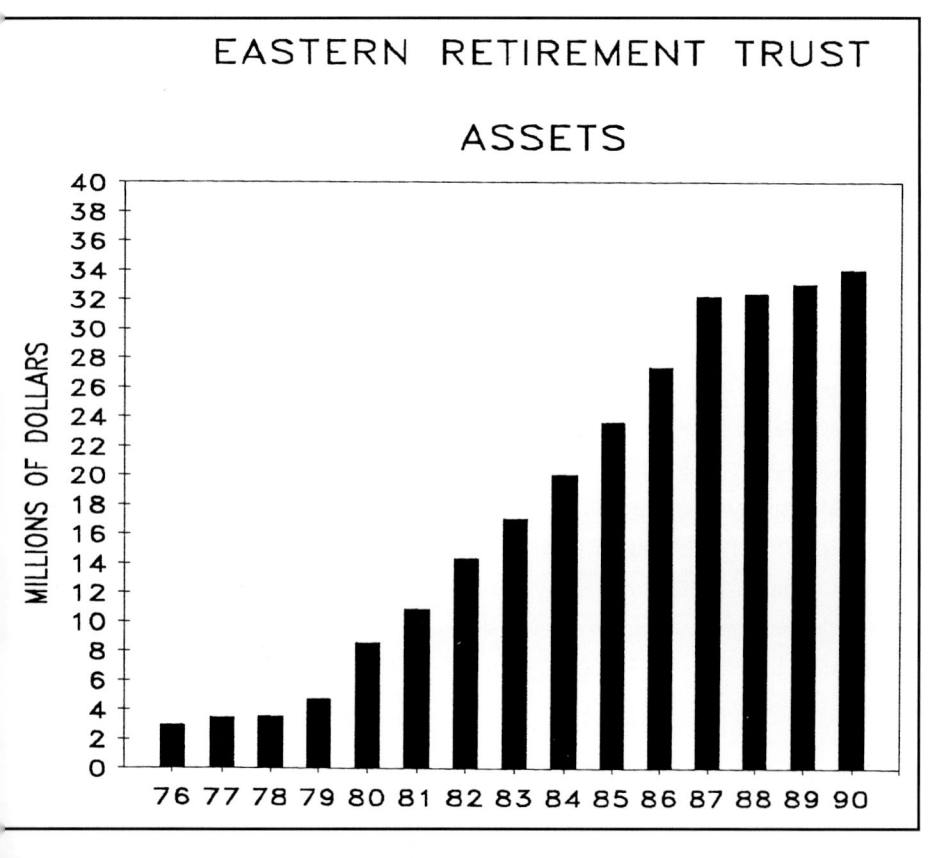

Opportunities to expand services provided by the corporation abound. On the horizon is the opportunity to help members meet the new requirements of the Environmental Protection Administration regarding underground storage tanks and the liability created under federal regulations. Such insurance opportunities are ample evidence of the value of a "member driven" organization.

MEMBER SERVICES

The Eastern Building Material Dealers Association offers a multitude of member services outside of the "traditional" big three: Group Insurance, Property/Casu-

association on behalf of the association program. However, they were to allow a participating member to select a local broker if he/she so desired.

During the next three years, the arrangements with Corroon & Black did not meet the expectations of the Mid-Atlantic Risk Management Committee. Because they failed to expand the program, the board authorized the investigation of alternative underwriters and managing brokers. The selection of Noyes Services in 1984 to perform the administrative functions provided the opportunity to use two underwriters in various aspects of the property casualty program. The PMA Group of insurance companies was chosen for the safety group in workers compensation coverage and commercial automobile coverage. Pennsylvania Lumbermen's Mutual Insurance Company was chosen for a special multi-peril program. The result of these changes was an immediate and dramatic upturn in participation.

During the six years that followed, participation more than doubled to 212 participating dealers and the number of insurance contracts expanded to more than 600. Annual premiums in the program now exceeds $8.8 million, roughly split between the two underwriters. Due to the skills of the underwriter, and the limitation of participation to employers who seek to have a better-than-average approach to safety, the success of the program has manifested itself in the form of substantial dividends—almost $1.5 million in dividends will be distributed to the participants for the year 1991.

In addition to the Property/Casualty Program, Mid-Atlantic Risk Management now is expanding its services into the creation of DOT Drug Testing Consortium for compliance with federal DOT regulations for truck drivers. In addition, additional policies are negotiated on behalf of members for integration with flexible benefit plan, 401(K) plan options and variable payroll deduction plans.

Opportunities to expand services provided by the corporation abound. On the horizon is the opportunity to help members meet the new requirements of the Environmental Protection Administration regarding underground storage tanks and the liability created under federal regulations. Such insurance opportunities are ample evidence of the value of a "member driven" organization.

MEMBER SERVICES

The Eastern Building Material Dealers Association offers a multitude of member services outside of the "traditional" big three: Group Insurance, Property/Casu-

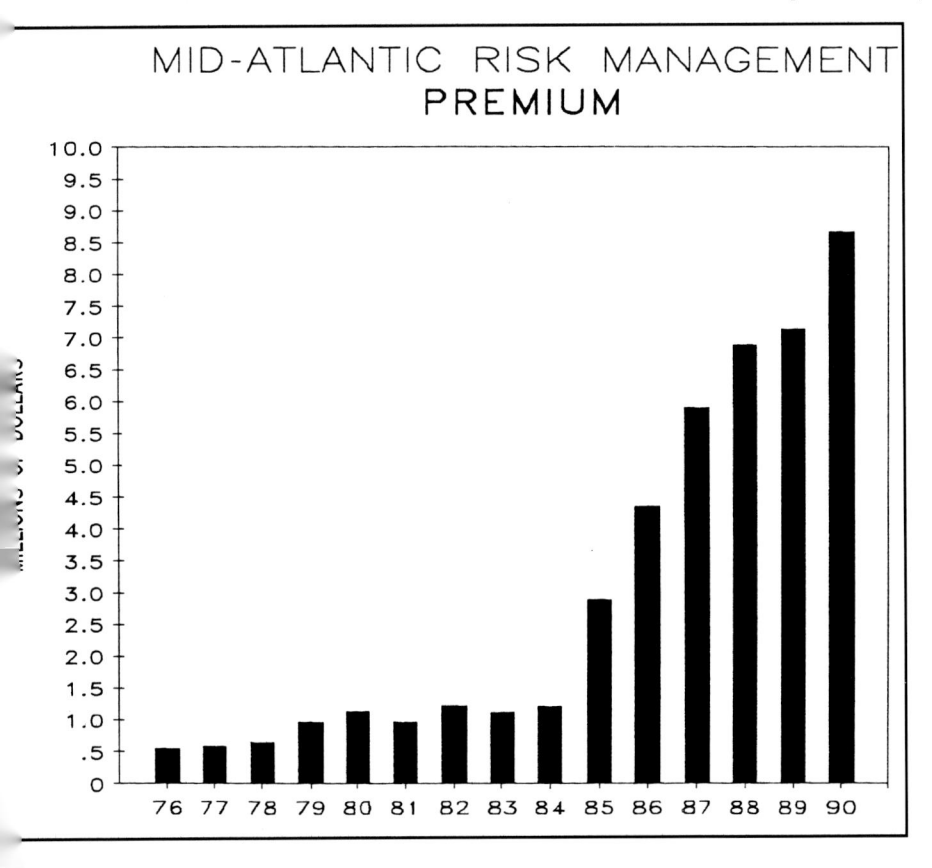

alty Insurance, and Retirement Trust Benefits. With the relationships developed over the years with various vendors, the association has been able to offer the member services that will have a favorable impact on their bottom-line expenses without having to go out to research and develop buying patterns of their own.

For example, Moore Business Forms have been affiliated with EBMDA for fifteen years. This is a relationship with the association that will provide savings on the purchase of business forms as well as other office stationery type products. The newest specialty of major computer software programs is the Matched Form Program, which allows Moore Business Forms to stock large quantities of preprinted forms that can be used on the major software programs that are utilized by the building material dealer.

Over the years, affiliations have been developed with various experts in specialty fields. This has resulted in an extensive list of consultants whose services can be offered to the Membership. These consulting services have either been used by the association, recommended and used by members, or developed exclusively for this industry. Consulting services that have been developed include actuarial, estate planning, freight bill auditing, polygraph, undercover operations and shopping services, wage and hour questions, data processing, financial and tax, legal, sales tax, security, and yard planning. As the need develops the consultants are geared to expand those consulting services being offered. Out of this list of consultants a loss prevention hotline has been established as has a legal assistance hotline. Members can contact experts in loss prevention and shrinkage and "employee-relations" attorneys who will answer specific brief questions from the members. The consulting service has even been expanded to include the management progress program, a consulting review service at the members' individual yard conducted by peer-dealers with similar problems and opportunities—at no charge to the member.

Over $3 million per year worth of credit card transactions go through the Visa/Mastercard Program. Again, this offers the majority of the membership the opportunity to use the credit card program at a reduced discount rate.

The Management Development Program is geared to promising young individuals selected by member firms for future management responsibilities. This allows the association to expose these young managers to learning experiences and social development within the industry.

By keeping extensive information on product information, the association provides help in locating distant manufacturers or manufacturers with an obscure trade name throughout the industry. This again saves the member valuable time by having the Association Membership Services Department conduct a product search.

With a variety of over sixty member services, the association has established a reputation in the development of member services that are bottom line oriented for the membership.

GOLDEN A
MIDDLE ATL
JANUARY 13 -

Golden Anniversary Convention, January 1942.
Photo courtesy of EBMDA

Chapter V
The Annual Meeting

Annual meetings play an important part in any organization such as this association. It is at the business sessions of these annual meetings where the business of the organization is officially transacted. In the early years of the association two meetings were held each year. The "annual meeting" was held in January and the records make it clear that it was this meeting where the important, official business was transacted. The second meeting was called the "semi-annual meeting" and, while some business was transacted, it was as much social as business and usually involved an outing to some local attraction.

The normal procedure for the annual meeting was to have a morning session where the various issues to be addressed would be raised and discussed. Committees would be appointed and the meeting would adjourn for a working lunch at which time the committees would draft reports. The committees would report at the afternoon session, which normally began at 2:00 p.m. These reports would set forth the association's position on issues, amend the bylaws, address membership concerns, and generally direct the Board of Directors to take certain actions. By 1905 they were even creating "Committees on Entertainment" for the meetings.

Souvenir Program with wooden covers for Young Men's Night, 1928 Annual Meeting (on following pages). Photo courtesy of EBMDA

The location of future annual and semi-annual meetings engendered considerable discussion at both meetings and at Board of Directors meetings. Eventually, this was addressed by delegating the responsibility for determining the site of the meetings to the Executive Committee. One gets the impression from the minutes that locations were generally decided upon based on the hometown of the president or other officers.

In the early years of the association, the annual meetings met in a variety of cities such as Allentown, Lancaster, Harrisburg, and Pottsville. However, by the turn of the century, Philadelphia had become the city of choice, due in part to ease of transportation, and also to the fact that merger talks were underway with the Philadelphia Association. Initially, when meeting in Philadelphia, the association met in the "Chambers" of the Philadelphia Lumber Exchange in the Bourse Building. Gradually, however, they moved to hotels, with the Bellevue Stratford becoming the hotel of choice.

The semi-annual meetings, however, were another matter, with the site generally being chosen because of some local attraction. Apparently the semi-annual meeting was more family-oriented, for references are made to activities for wives at the summer meeting but there is little reference to this kind of activity at the January meeting. Early sites of the semi-annual meetings included Reading, Gettysburg, Scranton, Williamsport, Easton, Mont Alto, and Wilkes-Barre. In 1906, the meeting was held in Chambersburg with a hotel rate of $2.00 a night and an excursion to the Gettysburg Battlefield available for an addition $1.55 per person (round trip). While the men were meeting in their business session, the women were given a tour of the local sanitarium.

When the semi-annual meeting was held in Lancaster in 1908, it was reported to the meeting that the Mayor had ordered that "the latch string was outside, and that the city belongs to the Lumber Dealers; that no arrests were in order and hoped that the dealers would return to their homes with not only a vision that Lancaster was a great city, but that each one would make a contribution towards a greater Lancaster."

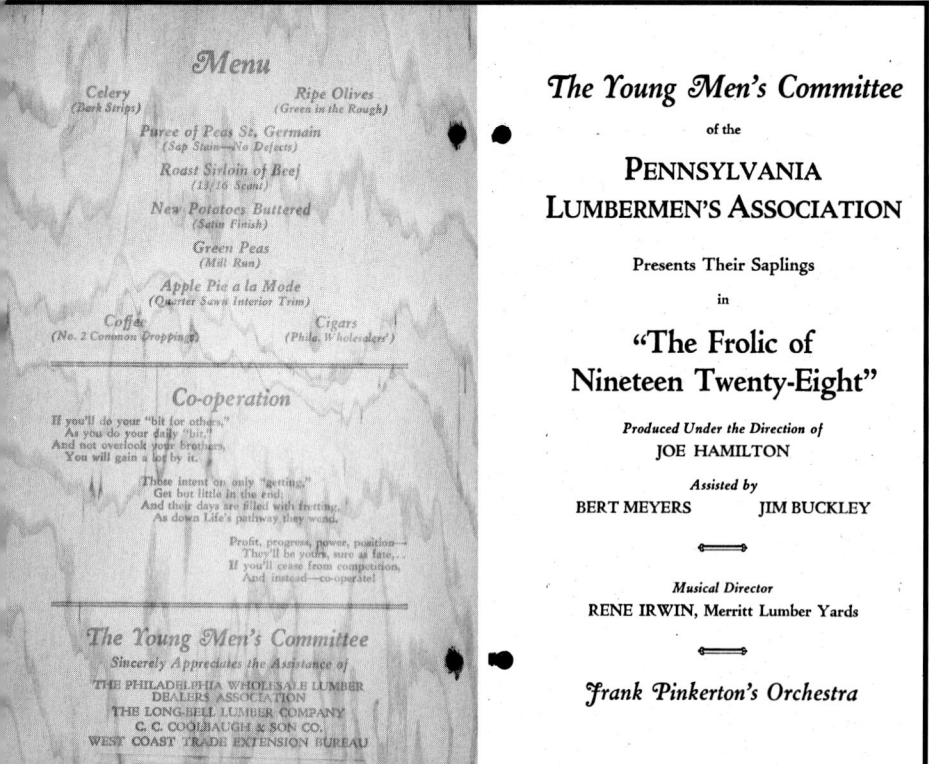

By 1908, the traditional pattern to the meetings, which had been in use since the beginning of the association, began to come under attack. At the 1908 Annual Meeting a resolution "urging a more interesting program be prepared for future meetings" was passed by an overwhelming majority. By the next year, the Executive Committee proposed that, beginning in 1910, a three-day session for the annual meeting be held, and the Bellevue Stratford was designated as the headquarters hotel. Included in the schedule of events for the first time was an hour for "Hoo Hoo," one of the few times that this social organization was mentioned in the records. The three-day annual meeting proved too ambitious and two-day meetings became the norm until more recent times. As part of the revamping of the annual meeting, the Executive Committee approved the payment of $23.00 for novelty buttons and $50.00 for "quartettes" for the 1910 Annual Meeting. In 1912, in an effort to boost attendance, members of the association were given free tickets for the annual dinner. Additional tickets cost $2.00 per person.

The year 1909 also saw the first efforts by representatives from Atlantic City, New Jersey, to have the semi-annual meeting held at that resort. However there was considerable resistance to holding the meeting outside of Pennsylvania, and the association decided on Scranton for their next semi-annual meeting. However, the opposition to going outside of Pennsylvania died down and in 1913 the semi-annual meeting was held in Atlantic City for the first time. The group was addressed by the mayor of the city: "We the people of

In The Show

Interlocutor
BERT MEYERS
Derr Lumber Company

End Men
BILLY DAVIS — Miller-Robinson Co.
ROLAND CONN — Derr Lumber Company

End Men
W. K. WALDIS — H. B. Wilgus Lumber Co.
JOE HAMILTON — C. M. Buzby & Sons

In the Circle

GEORGE BAUMANN	Chas. F. Felin Co.
EDWARD MOGCK	C. B. Coles & Sons Co.
WILLIAM T. HAMILTON	Wm. M. Lloyd Co.
H. C. REBER	Chas. F. Felin Co.
J. T. MURPHY	Chas. F. Felin Co.
ALBERT WILLIAMS	C. B. Coles & Sons Co.
RUSSELL L. GRAY	Derr Lumber Company
THOS. E. GIBBONS	Derr Lumber Company
ROY McDERMOTT	Wm. M. Lloyd Co.
GEORGE L. SHUTE	J. Gibson McIlvain Co.
JACK GALLAGHER	Hall Bros. & Wood
LEON FRETZ	Hall Bros. & Wood
CHAS. BRENNAN	Hall Bros. & Wood
J. E. SMITH, JR.	J. E. Smith & Sons
ALFRED ROBINSON	Derr Lumber Co.

Songbirds

BOB SCOTT	The Gillingham Co.
JOE SCOTT	The Gillingham Co.
BOB KISSINGER	Wilson H. Lear Lumber Co.
LES LA MAR	North Phila. Lumber & Coal Co.
BILL BARNES	Paramount Lumber Co.
DAVE BARNES	Shull Lumber Co.
A. CLYMER	Chas. F. Felin Co.
W. C. SMITH	Chas. F. Felin Co.
F. WHITTLE	Chas. F. Felin Co.
R. J. McLAUGHLIN	Chas. F. Felin Co.
C. BISSEY	Chas. F. Felin Co.
C. R. SECHLER	Chas. F. Felin Co.
E. P. McANDREW	Chas. F. Felin Co.
ALFRED C. MARGERUM	Chas. F. Felin Co.
JOS. J. SPONFELLER	Chas. F. Felin Co.
ALBERT H. TROTTNOW	Chas. F. Felin Co.
J. W. MARTINDALE	J. N. Arbuckle Co.
J. R. ROWLAND	Schively, Inc.
E. N. TAYLOR	Schively, Inc.
WILLIAM BOSTON	C. B. Coles & Sons Co.
LESTER MOUNT	C. B. Coles & Sons Co.

Program

Overture
Opening Ensemble ... ENTIRE COMPANY
Welcome to Young Men's Night ... BERT MEYERS
"Plantation Days" ... ENTIRE COMPANY
  (A) "HARVEST MOON"
  (B) "SILVER MOON"
  (C) "LINDY"
Introducing ... THE SCOTT BROTHERS
"Old Black Joe" ... JOE HAMILTON
"The Ghost of Old Black Joe" ... ENTIRE COMPANY
"Hello Folks" ... ENTIRE COMPANY
Tambourine Novelty ... ENTIRE COMPANY
"Halleluiah" ... END MEN
Chatter ... JOE HAMILTON
"Broken Hearted" ... RUSSELL GREY
"Where the River Shannon Froze" ... BILL WALDIS
Chips ... ROLAND CONN
Dance ... LEON FRETZ AND CHARLIE BRENNAN
Dollars and Sense ... ENDS
"Baby Your Mother" ... ROLAND CONN
Culls ... BILLY DAVIS
  Introducing "I WANT MY RIB"
"Longing" ... LEON FRETZ
Tongue and Groove ... BILL WALDIS
"Blue Heaven" ... BERT WILLIAMS
Jambs ... ENDS
"Oh, My Operation!" ... JOE HAMILTON
Introducing Four Old Friends in "Sawdust"
  BILL HAMILTON, RUSS ROWLAND,
  ROY McDERMOTT, AL ROBINSON
"In the Garden of Tomorrow" ... BERT MEYERS
Slippery Elm ... SCOTT BROTHERS
"Goin' Home" ... ED SMITH
The 2 x 4 in "Mixed Car Numbers"
  DAVE BARNES, LES LA MAR,
  BILL BARNES, BOB KISSINGER
"So Long, We'll See You Again" ... ENTIRE COMPANY
Finale ... ENTIRE COMPANY

THE trade-mark "Long-Bell" is known universally as signifying ideals and standards unsurpassed in the lumber industry.

The following products bear this trade-mark:

Douglas Fir Lumber, Timbers, Door and Window Frames; Western Hemlock Lumber; Western Red Cedar Siding and Shingles; Southern Pine Lumber and Timbers; Southern Hardwood Lumber and Timbers; Oak Flooring; California White Pine Lumber, Sash and Doors, Box Shooks; Creosoted Lumber, Timbers, Posts, Poles, Ties, Guard-Rail Posts, Piling.

The Long-Bell Lumber Company
R. A. LONG BUILDING   Lumbermen since 1875   KANSAS CITY, MO.
PHILADELPHIA SALES OFFICE - - 1413 PENNSYLVANIA BLDG.

1928 P. L. A. CONVENTION

May the memories of this Young Men's Night linger long as having been one of real fun and enjoyment for you. And may you continue to smile until a year hence. Then—come again!

C. C. Coolbaugh & Son Co.
Strictly Wholesale Millwork

19th & Cambria Sts.
Philadelphia
DIAMOND 6401

Hudson St. & Boulevard
Gloucester City, N. J.
GLOUCESTER 402-3-4-5

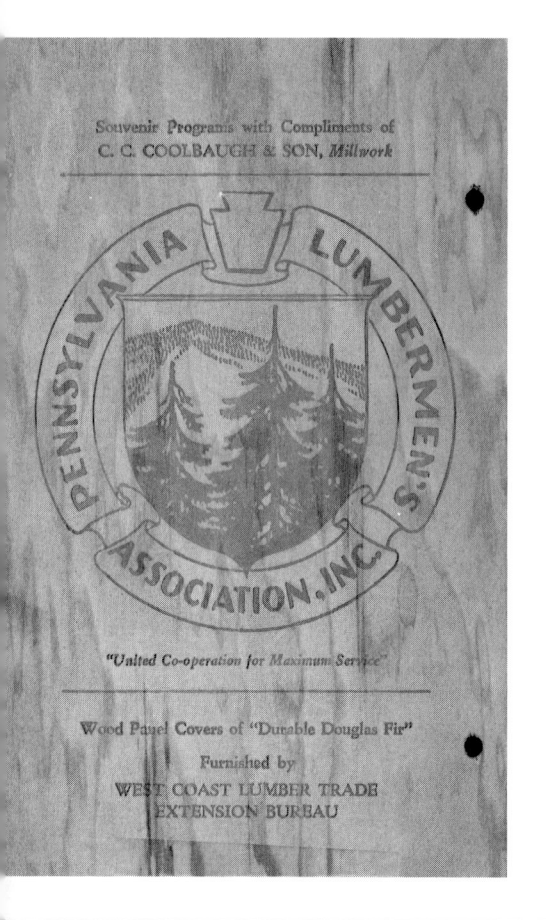

Atlantic City are on our knees to you visitors for coming here and spending money. The merchants and hotel keepers reap their golden harvests from you, and your kind. Go out and have a good time. We have just granted 185 licenses, and that ought to be a preventive from your going thirsty." At the end of the meeting the association voted a resolution of thanks to Mayor Riddle for "his most cordial welcome and suggestive address."

As the decade progressed, there would be a number of revisions in the annual meeting to reflect changing needs. In an attempt to increase membership outside of their normal area of Pennsylvania, annual and semi-annual meetings were held in Wilmington, Delaware, Ocean City, New Jersey, Bedford Springs, Pennsylvania, with return visits to Atlantic City, and to resort areas in the Poconos like Delaware Water Gap as well.

The content of the meetings also changed. This was most noticeable during the period after 1915. Prior to this time, most sessions dealt with the specifics of the lumber industry and only occasionally were there sessions on more general topics. Increasingly, however, sessions dealing with such topics as credit operations and insurance were held. There was also increased interest in providing entertainment and a nationally prominent speaker for the annual dinner. At the 1916 meeting, held in Philadelphia, the members of the association were invited by John Wanamaker to tour his store. The tour included "a trip to the roof, the cold storage vaults, etc., and an opportunity to hear the Great Organ, the largest in the world."

The war years, in particular, saw changes in the format of the sessions. Scheduled were general information topics such as "Thrift and War Savings Stamps" and the "The Income Tax," conducted by John F. McEvoy, chief clerk, Philadelphia Internal Revenue Office. During and immediately after the war, sessions reporting on American soldiers and their activities were held. The 1919 meeting included a session conducted by the Bureau of Public Health, Surgeon General's Office, on "Venereal Diseases Affecting Industrial Establishments."

Probably the most important development affecting the annual meeting was a vote taken at the 1919 meeting "to allow the Johns-Manville Company to have an exhibit of their wares near the meeting room." This vote would change dramatically the nature and content of the annual meetings. By 1923, the association required an "exhibition room" as part of the annual meeting, and entertainment became an important part of the program. The 1922 annual meeting had a budget of $300 for speakers and $120 for an orchestra for

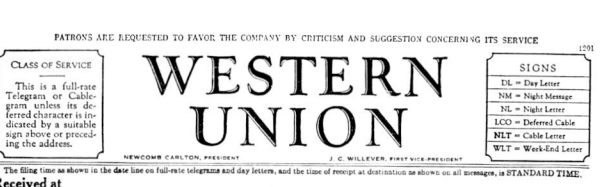

Telegram from President Herbert Hoover for the Thirty-ninth Annual Meeting, January 1931. Photo courtesy of EBMDA

the dinner. Dinner speakers were generally of national importance. While they were not able to get Vice President Calvin Coolidge or John D. Rockefeller as speakers, they were able to attract Secretary of Agriculture Henry Wallace in 1924.

By the mid-1930s efforts were underway to again change the format of the annual meeting. Shortly after the 1935 meeting the Board of Directors agreed that something had to be done to improve attendance at the business sessions. In the words of board member Fred A. Ludwig, "something should be done to bring these dealers into the meetings as the Convention is primarily arranged for educational purposes, with fine speakers being paid honorariums to bring the latest methods and views on the most vital questions of our businesses; whereas the Convention seems to be a hand shaking affair centering around the exhibitors for a good time and a drunken orgy." The Board (members) were quite in accord with Mr. Ludwig's views and felt that some definite steps should be taken to overcome this situation. The issue was placed before the Arrangements Committee, who decided it was too late to make any major changes in the program in 1936. However, some minor alterations did take place:

> . . . door prizes are to be given out at each session amounting to $25 each at each session. Slips will be passed out before 2:15 P.M. All holders of lucky numbers must be in the business session at the close in order to receive a prize. Exhibitors, manufacturers and wholesalers are to be written that the Association frowns on the practice of serving drinks in rooms and will consider it a discourtesy if this is done. It has gotten to be a matter of competition to see which wholesaler can set up the best bar and this is reacting adversely on the convention in driving dealers away. The Bellevue is also to be told of this. It was suggested that the Warwick Hotel and the Penn Athletic Club might be used in the future if the Bellevue is not cooperative.

That year there was still a Stag Party on the last night of the convention—a wrestling match in Camden. This Stag Night was in addition to having Robert Fleming, president of the American Bankers Association, speak the second afternoon of the convention. General Hugh S. Johnson, the former head of the now defunct National Recovery Administration, was paid the grand sum of $500 to be the opening speaker on the first

Fashion show at the 1953 Annual Meeting. Hunt Bowers, Art Hood, and Bill Costello are the men trying to place the hats.
Photo courtesy of EBMDA

Charlie Graff and his wife Jessie at the 1953 Convention.
Photo courtesy of EBMDA

Business meeting at the 1958 Convention.
Photo courtesy of EBMDA

Past President Pat Malone and his wife, Betty, at the 1958 Convention.
Photo courtesy of EBMDA

Past Presidents Hunt Bowers and Bill Wolf at the 1958 Convention.
Photo courtesy of EBMDA

day of the convention. This was all organized under the theme for 1936, the "Fight Year," as the association emphasized that the chief business of the dealer member should be making more prosperous customers, not new stock and new yards.

In 1937 the meeting was changed to a two-day affair to which wives were invited. The "Young Men's Night" was changed to "Frolic Night" and ladies were invited to participate. Also, an effort was made to obtain the Unviersity of Pennsylvania Glee Club for entertainment. It was eventually reported that activities for the women went over well and that the women in attendance organized their own committee to provide activities for women at subsequent meetings. It was decided to solicit the home address of the head of each dealer member, and their wives were to be sent an invitation to attend the convention. By 1940, the board decided to go back to a "Men's Night," but there was also a dinner and a theater for those who brought their wives.

Other efforts were undertaken to make the annual meetings more attractive and thus boost attendance. Joseph W. Brosius, president of MALA, 1938–1939, remembers a particularly effective approach for getting a bigger crowd to the annual banquets:

> The annual banquet in 1937–1938 had been very poorly attended and Bob Jones decided to do something about that. He originated or adopted the idea of granting an award to a speaker who might draw a crowd. During the operation for the 1939 Annual Meeting, Bob wrote a letter to a Mr. Brown. I believe he was President of Johns-Manville Corporation. This was in late 1938. He informed Mr. Brown that the Middle Atlantic Lumbermens Association had awarded him and his company their highest honor for outstanding service given to the lumber industry and wished to present him with a certificate and plaque. Would he attend our February 1939 Annual Meeting to accept the plaque and be our principal speaker? Mr. Brown was very prompt in accepting, and when the time came, he was on hand with his PR and all his entourage. In February of 1939, a Johns-Manville salesman in our area urged dealers to come to the meeting to hear their President. Everybody must come to hear him speak. The night arrived, the Ballroom was packed. Mr. Brown accepted the plaque and was the principal speaker. I was impressed by the fact that in the morning of the 29th, I was called to appear at a certain suite bringing the plaque to make a presentation to Mr. Brown before the

press photographers. I rounded up George Kingsley, who was Vice President of the Association, and we went to the appointed place where we found everyone waiting and ready. We stood one on either side of him and I presented him with the plaque. I was told that the picture was printed in the evening paper in San Francisco that night. The practice was continued after this success.

Throughout the 1920s and 1930s and early in the 1940s the preferred site for the annual meeting was Philadelphia and, in particular, the Bellevue Stratford Hotel. During the 1940s the preference seems to have shifted to Atlantic City, particularly the Chalfonte-Haddon Hall Hotel. By the 1960s there developed a conscious effort to move the annual meeting around to various sites within the geographical region of the association. Increasingly, however, it became a problem finding a hotel big enough to hold the event. More recently, the annual meeting has become a building materials exposition for "The Eastern Market," and the size of the event has mushroomed far beyond anything originally dreamed of when, in 1919, the Board of Directors voted to allow Johns-Manville to have a display outside the room where the annual meeting was held.

Bob and Marion Jones at the 1958 Convention.
Photo courtesy of EBMDA

Luncheon and fashion show for wives at the 1958 Convention.
Photo courtesy of EBMDA

1976 Convention. Front row (left to right): Ray Bures, Dave Charron, Bob Wood. Back row (left to right): Russ Allen, Edgar Harman, Linc Dillingham, Bob Moore, Harry Johnson.
Photo courtesy of EBMDA

Diamond Jubilee Convention Program, February 1967.
Photo courtesy of Smick Lumber and Building Materials Center

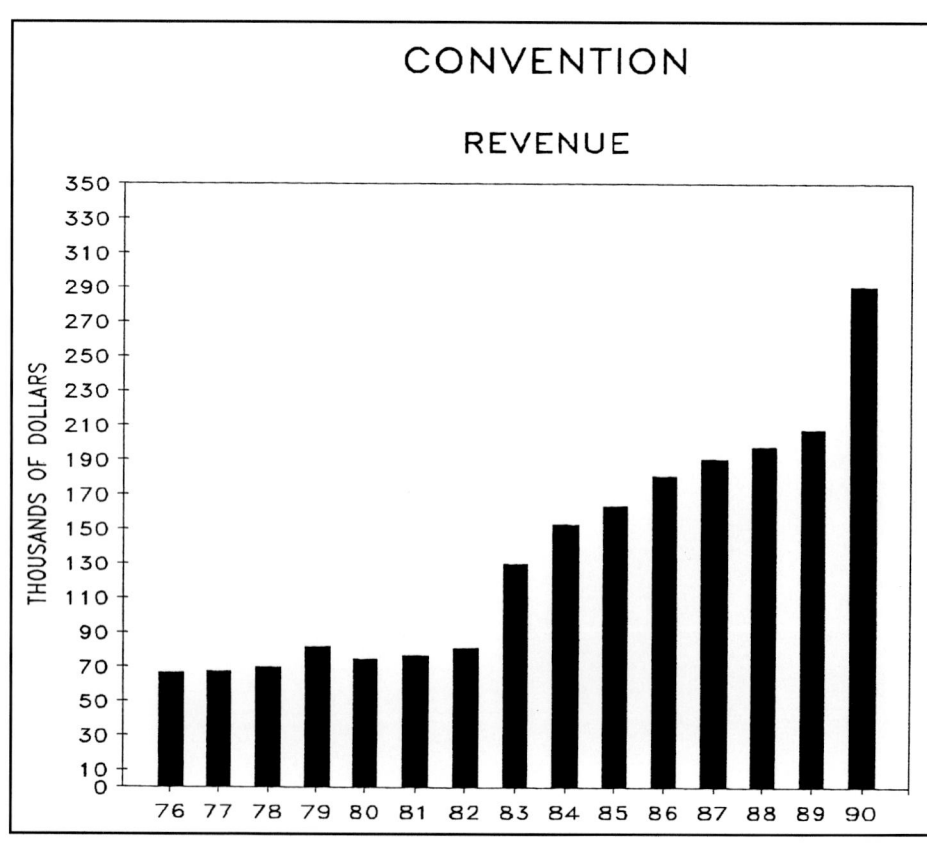

Past President Richard A. Kauffman and Executive Vice President Russell J. Allen at the 1975 Convention.
Photo courtesy of EBMDA

(below) Centennial Convention Advisory Committee Top Row (left to right): David G. Patterson; David B. Kreidler; Bernard Bernstein; John H. Eaton, Jr.; William E. Baer, Jr.; Gene S. DiMedio; Edwin F. Scholtz; and Frank Kelly. Middle Row (left to right): John F. Beaver; James C. Krebs; William P. Wallace; Bruce C. Ferretti; Steven Boyd; G. Eugene Mackie; Dale C. Adams; and James P. Rauch. Bottom Row (left to right): W. Russell Lamar, Jr.; Raymond Weitzel; J. Fred Robinson; Vincent J. Tague; Joseph C. Bradley, Jr.; Richard W. Brown; and Harry H. Johnson III.
Photo courtesy of EBMDA

The 1963 fire at Isaac Smick and Son Lumber Company also destroyed a portion of the town of Quinton, New Jersey. Photo courtesy of Smick Lumber and Building Materials Center

Chapter VI
Insurance

As indicated earlier in this work, insurance was one of the first concerns of the association. The subject was first broached at the semi-annual meeting in July 1894, to investigate the "feasibility of creating an insurance company for Lumber Dealers within the Association because of the high premiums paid by dealers in the state." This Committee reported to the Annual Meeting in January 1895, that they had obtained information from fifty-three firms that had paid insurance premiums of $244,400 over a ten-year period for $13,210,000 in insurance protection and had received only $17,446 in payment for losses. The Committee reported that a mutual insurance company was feasible and that the only question that remained was how to proceed. The annual meeting called for the creation of a committee to work with the Philadelphia Lumber Exchange "to create such an insurance company for lumber yards, woodworking establishments and members of different lumber associations."

By the semi-annual meeting in July 1895, the company, Pennsylvania Lumbermen's Mutual Fire Insurance Company, was in operation. One of the topics discussed at the 1895 meeting was whether the company's future growth would be retarded if membership was limited to members of the association. By January 1896, the Committee on Insurance reported that the company had been created; that it had written $1,000,000 worth of

> **WHY**
>
> **The Pennsylvania Lumbermens Mutual Fire Insurance Company**
>
> WAS ORGANIZED:
>
> **I. Because statistics show that the stock companies have been overcharging our trade for years.**
>
> The following figures are from Whiting's Insurance Directory for 1894. Compare the premiums collected from the wood-working industries with those collected from the textile industries, and notice the proportion of loss in each line. The low rates obtained by the textile industries are due to the fact that they have competed with the stock companies by independent insurance movements.

Advertisement for the Pennsylvania Lumbermen's Mutual Fire Company in the 1895–96 "Constitution and By-Laws."
Photo courtesy of EBMDA

insurance with premium payments of $14,000; and that there had been no losses the previous year. While a detailed history of the insurance company is outside the scope of this work, it should be pointed out that a long and fruitful relationship existed throughout the history of both organizations. In the early years, officers of the association served on the board of the insurance company, and one of the highlights of the early annual meetings of the association was the reports from the president of the insurance company on the success of this endeavor. The insurance company was repeatedly pointed to as an example of what united action could accomplish.

This connection between the insurance company and the association also provided an extra bonus for the association: the insurance company provided a set of brass quoits for presentation to the winner of the annual quoit competition held at each semi-annual meeting.

Gradually, the association and Pennsylvania Lumbermen's Mutual went their separate ways, although it is apparent that Pennsylvania Lumbermen's Mutual insured almost all of the members of the association for fire insurance. There were two periods when friction developed between the two organizations. First, in the early 1920s there were charges made at some association meetings that the insurance company was charging excessively high rates for their fire insurance. However, this was investigated and satisfactory explanations were presented to the association. In the 1930s, as Pennsylvania Lumbermen's Mutual attempted to diversify its base of operation and branch out into non-lumber industry areas, charges of excessively high rates were once again raised by the association. In fact, the association led the way in providing members with alternative fire insurance carriers as a way of assuring competition. After a delegation of association officials met with the Board of Directors of the insurance company, an amicable settlement was arranged and the premiums were reduced to a more competitive level. This close working relationship between the association and the Pennsylvania Lumbermen's Mutual continues today as the company both provides insurance coverage for EBMDA members and also acts as underwriters for some of the other association's insurance programs.

For almost half of the association's history, the only insurance provided was through the separate operation of Pennsylvania Lumbermen's Mutual. Beginning at the end of the 1930s and continuing to the present, providing an increasing variety of insurance programs has become a major part of the association's programs for members. In October 1938, the Board of Directors agreed to "establish an Insurance Service for the Association." Shortly thereafter, it was reported that the service was established and functioning. Details of the this insurance service were never presented in the minutes of the Board of Directors, so it is difficult to determine the exact nature of the operation. But within a year, the board was informed that the service had delivered seven policies worth $240,500 and that there had been considerable savings. Furthermore, sixty-two brokers of record had been filed. Apparently, as part of this increasing emphasis on providing insurance programs, in July 1939, the association decided that it should "commercial as many services as possible" in an effort to maximize income. Fred Martin and Bob Jones were instructed "to obtain insurance brokers licenses and collect a percentage of the broker's fees that might be used to pay a portion of their salaries." By the end of the next year, the association treasurer, Ray Latshaw, reported that this new insurance service had provided forty-four dealers with policies and that the association had received in excess of $2,300 in commissions.

Approximately ten years later, on September 24, 1948, the association took another giant step in insurance services when the board approved the following resolution to establish a pension plan for association members:

> Resolved, that the Pension Plan Committee be and is hereby authorized to give its final approval to the Trust Agreement and Pension Plan as prepared by Saul, Ewing, Remick and Saul, Esq., and Ostheimer and Company and approve the Pennsylvania Company for Banking and Trusts as Trustee and to authorize Ostheimer and Company to solicit participation in the Plan by the individual members of the Association and to take all action which may in its discretion consider necessary to place into operation and carry out the purposes of Middle Atlantic Lumbermens Association Pension Plan.

Three months later, on December 13, 1948, the Board of Directors agreed to set up the Middle Atlantic Lumbermens Association Group Insurance Trust at a cost not to exceed $1,500. The secretary of MALA was to act as secretary of the trustees, and the board elected three trustees from among their ranks. This was to be the beginning of the first medical/health insurance program offered by the association.

By March 1949, it was reported that both the Pension Plan and the Group Insurance Plan were operational and that they would have the requisite number of participants to begin operation on May 1, 1949. This apparently happened because at the June 1949 meeting of the Board of Directors it was agreed to have MALA employees covered by the Pension and Group Insurance Plan. It is interesting to take note of the first report on the activities of the Group Insurance Trust. At the September 15, 1949 meeting of the Board of Directors, the Group Insurance Trust reported on their first nine months of operation and the figures are given below. For

Scenes of the 1965 fire at the Mizell Lumber Company.
Photos courtesy of Mizell Lumber and Hardware Company, Inc.

131

comparative purposes, the figures for forty years later are also given.

	1949	1989
Members participating:	25	273
Employees/dependents insured:	532	5,289
Premiums paid:	$5,414	$9,964,706
Claims paid:	$1,746	$7,231,197

In 1949, the total cost of hospitalization was $441.60 and the total surgical expenses were $300.

By December 1950, the Group Insurance Trust had been successful enough that the association could recover the initial start-up costs that they had advanced to the trustees and they could begin to collect the current charges for administering the program. As a result, the trust was charged 50 percent of the cost of one secretary's salary, one-third of the salary of Ray Latshaw, the treasurer of the association, and one-half of his travel expenses. By this time, almost nine hundred employees and dependents from more than sixty member firms were covered by the program. By the late 1960s, the Group Insurance Trust added a major medical plan, a profit sharing plan, and expanded the pension plan. Their operations proved so successful that other lumber associations, such as the West Virginia Lumbermens Association, had them administer their medical programs. By the end of the decade, MALA was able to brag that their Group Insurance Program was the largest of all the federated groups within the National Lumber and Building Materials Dealers Association.

At this time, a reorganization of the Pension Plan took place. It is difficult to trace through the minutes of the Board of Directors; however, former President Joseph Haenn, Jr., recalls:

> In 1969 we had the Association (a tax free group because it was educational), and we formed MALA Inc. (a tax paying division that handled things doing with money received and spent), and a toddling group called the Group Insurance and Pension Plan Trust. This administered the various health and welfare and retirement plans that were offered our dealers. There was no large influx of money, though. The medical plans were strong but in a growing stage. We had just started to handle plans for the West Virginia Lumber Dealers Association. Our retirement programs were in bad shape. In fact, the bank then handling our fledgling pension plan made some bad calls and the Pension Plan lost money in 1969. Luckily, it hurt Cliff Cramer's R. C. Cramer Lumber Company, East Stroudsburg, Pa; and my company the most. We were both

The 1963 fire at Isaac Smick and Son Lumber Company also destroyed a portion of the town of Quinton, New Jersey. Photos courtesy of Smick Lumber and Building Materials Center

officers of the Association, and both believed the Plan would work if it were allowed to get a little stronger. We both stayed with it, and I don't remember any other participant being scared enough to withdraw. The bank holding the Pension Plan was changed, though.

The structure of the Group Trust and Pension Plan was also changed shortly thereafter, in 1973, with one Board of Trustees being established for the Retirement Plan and a separate Board for the Group Insurance Trust. Fortunately, by this time, the Retirement Trust had grown to approximately $1,750,000. Today, the Retirement Trust has 101 participating plans with assets in excess of $36 million with more than $4.4 million being paid out last year in benefits.

The insurance operation of the association would be expanded in December 1978, when a wholly-owned subsidiary corporation, Mid-Atlantic Risk Management, Inc. (MARM), was created to serve as an insurance agency subsidiary. The officers and Board of Directors of MALA, Inc., were the officers and Board of MARM. In the articles of incorporation, MARM's purpose was listed as:

> engaging in the business of insurance by acting as agent for insurance companies in soliciting and receiving applications for general life, health, accident, casualty, liability and all other kinds of insurance, collecting premiums, and offering such other businesses as may be delegated to agents by such companies, and conducting a general insurance agency and insurance brokerage business, and doing of all things necessary, customary or incident thereto, and to do all things and exercise all powers, rights and privileges which a business corporation may now or hereafter be organized or authorized to do or to exercise under the Business Corporation Law of the Commonwealth of Pennsylvania.

At their first organizational meeting, the Board of Directors of MARM authorized the officers of the corporation to apply for a brokerage license as well as for insurance agent's licenses for life, accident, health, and casualty. They also indicated that while they would concentrate on providing insurance coverage for lumber dealers, they would provide insurance plans for other industries as other appropriate opportunities existed.

With this reorganization and the creation of MARM, the association was now able to provide a complete range of insurance programs for members of the Eastern Building Material Dealers Association.

THE PLAN

Bruce H. Helfrich, Geo. Helfrich & Sons, Baltimore, Md., 19th President of the Pennsylvania Lumbermen's Association

FEBRUARY, 1933

In This Issue
Complete Resume of the Convention and Other Features

Chapter VII
Publications

From the very beginning of the association, publications always played a very important role in its success. Within a year of the founding, the Pennsylvania Lumbermen's Protective Association felt the need to have an "official organ" and designated the *New York Lumber Trade Journal* as their official publication. In return for a fee of $1.00 per member, the association was able to send material for inclusion in the *Journal*.

As has been pointed out earlier, one of the mysteries of the early association was how they communicated with members. It is clear from the records that some type of publication was being sent to members. Whether this was a formal newsletter, a circular letter, or some other type of publication remains unclear. However, we do know that they published some type of "report," if for no other reason that to list those companies which had been censured by the association and to publicize those retail lumbermen who were members. There is also evidence that the association printed some type of list of lumber retailers in the state for the use of wholesalers. Furthermore, there is evidence cited earlier in this work that in 1898, the secretary of the association wrote to members who had not paid their dues informing them that

---

Photo courtesy of EBMDA

> How to Take
> Advantage of the
> Mechanics' Lien Law
>
> IN
>
> *Pennsylvania*
> *New Jersey*
> *Delaware*
> *Maryland*
>
> Compiled by and Prepared
> for the Members of
>
> The Pennsylvania
> Lumbermen's Association
> 212 Otis Bldg.
> Philadelphia

Pamphlet issued by PLA in late 1920s providing information to dealer-members on mechanics' lien laws in the various states served by the association.
Photo courtesy of EBMDA

for the past year and a half the Secretary . . . has mailed to your address our regular reports relating to the abuses practiced by said wholesale shippers hoping that the reports would be appreciated by you to the extent of your taking an interest in the workings of the Association and becoming a member thereby aiding both yourself and the retail trade generally in eradicating the evils which the trade is subject to.

Apparently the association attempted to have a joint monthly "Report" printed that would include the businesses being listed by the New York, New Jersey, Massachusetts, Connecticut, and Pennsylvania associations. The effort was to reduce printing costs and, at the same time, to provide for a wider distribution of the names of malefactors as determined by the associations. Whether this was ever implemented is a mystery.

The only publication from this early period that is still extant is The Constitution and By-Laws and List of Members of The Pennsylvania Lumbermen's Protective Association, which was printed 1895–96 and which is a handsomely bound volume that clearly was to be used not only as reference, but also as an advertisement for the association. It included the "Constitution and By-Laws", a list of all members, including honorary members, pictures of all the officers, and a listing of all the committees as well as advertisement's from a variety of lumber dealers around the state.

The methods established for this early publication set a pattern for most publications of the association—generally based on some type of private entrepreneurship and self-sustaining and profit-making, if possible. The book was given to a Mr. D. J. Keith to produce and he reported in July 1895, that the total expenses for publishing the reference book was $1,780.88, providing a profit of $1,329.12, of which the association received 50 percent or $759.56.

After this, there are scattered references to publishing efforts on behalf of the association, but little description is provided. We know that by 1902, the other lumbermen's associations in the Eastern Association were suggesting that each of the constituent groups publish a list of members "like the Pennsylvania Lumbermen's Protective Association does."

By 1910, the advisability of publishing an association magazine had considerable support within the Board of Directors. *The Plan* was launched, and by July 1911, the secretary and a Mr. Joseph W. Tatum, of the Biddle Press, were complimented on the success of the new journal, which apparently was published three times in 1911. The total cost of the three issues was $437.16. Of that cost, stamps and wrappers accounted for $33.66, and the actual printing cost was $378.50. Mr. Tatum received a commission of $25.00. This proved financially rewarding because advertisements brought in $518 for a profit to the association of $80.84.

Based on this initial success, the association entered into an agreement with Tatum and the Biddle Press on January 6, 1912, to publish *The Plan* as a monthly journal. The two-year contract called for the association to act as publisher and for Tatum and the Biddle Press to be the printers. Tatum agreed to print and distribute and bear all the expenses of such printing and distribution provided that proceeds of advertisements, which are at present carried in *The Plan*, and all future advertisements were turned over to his parties. In other words, in consideration of the revenue for advertising in *The Plan*, Mr. Tatum and his parties agree to conduct the publication of it under the supervision and control of the Association, free of any expense to the Association.

The annual subscription price was to be $1.00 that was to go to the association and "the top half of the inside front cover was reserved for a roster of officials without cost" to the association. Within a year, however, the agreement had to be modified to provide either for $300 in additional advertisements or have the association provide the difference in cash to cover the cost of procuring news and writing articles.

This arrangement was continually renewed until 1921 when the president and secretary were instructed to negotiate a new contract that would provide $1,000 to the association. Apparently, this was not mutually agreeable, because by the end of the year the Board of Directors were suggesting that Mr. Tatum be hired by the association as an assistant secretary to devote his full energies to *The Plan*. This, too, apparently did not work out, and in April 1922, the Board of Directors approved hiring J. H. Reiter, a student at Haverford College, as field secretary. He was to divide his time equally between *The Plan* and member recruitment. One half of his salary of $1,500 was to be charged against the magazine—he was "to take charge of *The Plan*." By the end of the year, the Board of Directors was discussing the advisability of separately incorporating the journal. It is obvious that its offices were located at the headquarters of the association and that the secretary, Fred Martin, was acting as treasurer of the journal. By the end of the year, the board agreed to hire additional clerical help for Mr. Reiter.

From the minutes of the Board of Directors, it is apparent that they were concerned about the health of *The Plan* and were continually having committees look into its operation and providing advice. However, under Reiter's guidance, the magazine apparently began to thrive, for he reported on increased advertising, and by 1923 he reported an income increase of 50 percent over the previous year. By May 1923 he was able to report that it cost $6,000 to produce nine hundred copies of the January 1923 issue and the same to produce thirteen hundred copies of the May issue. This amount included the salary of Reiter and a stenographer.

While Reiter was experiencing apparent success with the magazine, he

MALA issues annual surveys of business for dealer-members.
Photo courtesy of EBMDA

An example of publication used by the association to help dealer-members to develop their marketing and selling skills.
Photo courtesy of EBMDA

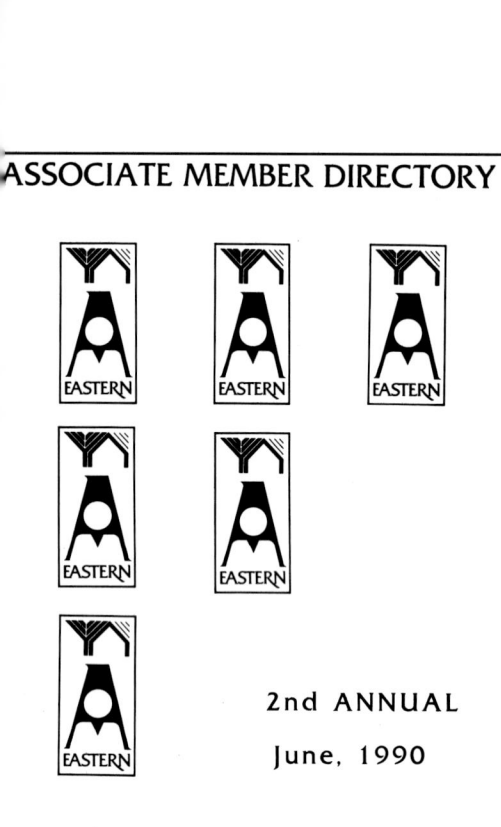

Photo courtesy of EBMDA

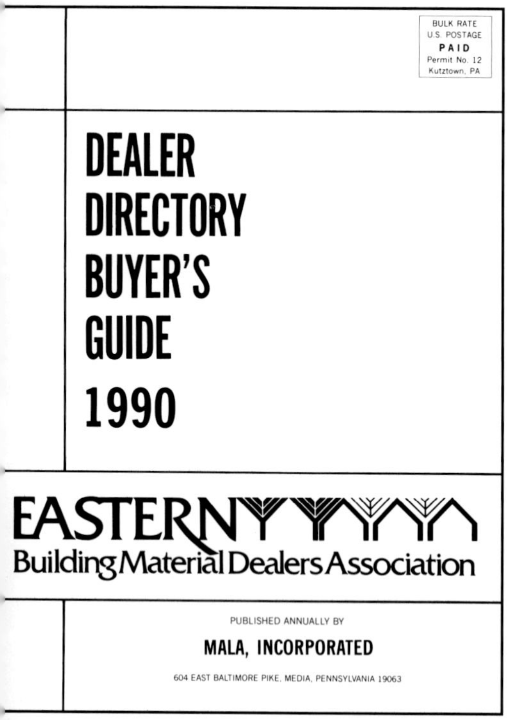

Photo courtesy of EBMDA

was not as successful in his efforts to recruit members, and he resigned as field secretary in September 1924. It is unclear from the minutes of the Board of Directors, at this point, exactly what happened to *The Plan* because there is no further mention of it. Whether it was separately incorporated, as suggested, and Reiter remained with the magazine is a mystery. We do know that it was successful because a profit of $1,000 was reported for 1924.

At the same time that these changes were taking place with *The Plan*, the association became involved with another publication, *The Successful Merchant*. This, apparently, was a national publication distributed to members through the auspices of the association. The Board of Directors had some early misgivings, especially over liability and whether the "plans" in the magazine were copyrighted. However, by 1924 these misgivings, were apparently clarified. The association participated in the distribution of the magazine, because it reported an income of $1,000 from it.

It was not until the meeting of the Board of Directors of March 1, 1935, that any publication was mentioned again. In the minutes of that meeting it is mentioned that a Dr. Hess of the University of Pennsylvania was hired to act in an advisory capacity to *The Plan* and that he was to write articles for the journal. After this cryptic note, nothing appeared about the magazine again until March 22, 1943, when the Executive Committee approved the submission of a reorganization plan. The Board of Directors then approved the reorganization as proposed by A. J. Fehrenbach and agreed to hire him as executive editor.

Fehrenbach was a 48-year-old graduate of the University of Wisconsin School of Journalism who had formerly worked for the *New York Times*. He had been an associate editor of a printing journal and an editor of a Girl Scout magazine. In presenting his reorganization plans, Fehrenbach told the board:

> *The Plan* is a powerful instrument. It has an honorable and respected record. It patiently needs careful streamlining and modernization. Without sacrificing any of its conservative and substantial values, its editorial caliber should be stepped up and the tensile strength of the ideas it advances should be increased. Given the proper thought and application to bring about the needed improvement the publication will have the stamina to carry the banner for its owner's traditional rights as good citizens and honest businessmen and to withstand the assaults of bureaucratic theories and practices that threaten to undermine all successful commercial enterprises.

His plans called for increasing the size of the magazine to sixty-eight pages with six pages of ads. He was to be employed for an experimental period of four months at a rate of $75.00 per week plus expenses. If both he and MALA were agreeable to the arrangement at the end of that period, he would be employed under the same conditions for the remaining four months of 1943. "If *The Plan* magazine for the year 1943 earns a net excess over the $3,000 as provided in the budget of MALA after all expenses have been charged to the magazine plus all payments made to Mr. Fehrenbach the entire excess over and above the $3,000 as budgeted shall be paid to Mr. Fehrenbach." After January 1, 1944, he was to be paid an annual salary of $4,000 and any net earnings remaining in the magazine above $3,000, after all expenses have been paid, were to be divided equally between MALA and Fehrenbach. This arrangement apparently worked out because in June 1944, Fehrenbach received $500 as his share of the 50 percent of net earnings. However, by October 1945, these arrangements were no longer satisfactory and Fehrenbach resigned effective December 31, 1945, because "he did not believe that it was possible for *The Plan* to provide the revenue the Association needs and he needs." The board accepted the resignation and gave Fehrenbach a leave of absence until December 31, 1945, with full salary. The board appointed another committee to review the operation of the magazine.

While these changes were taking place with *The Plan*, the association began another publication, the *Dealers Directory and Buyers Guide*, which continues to be published today. When initially printed in 1943, it was to be distributed yearly at the time of the annual convention. As with the other publications, it was expected that this new one would produce a profit for the association.

By the end of the year, yet another reorganization of *The Plan* occurred with the magazine being sold as a way of making MALA a non-profit association. A separate corporation, *The Plan*, Inc., was established and a value of $1,000 placed on the name and copyrights of the magazine. When the IRS granted tax-exempt status to the association, *The Plan*, Inc. was separately chartered on September 10, 1946. The executive director of MALA, Robert A. Jones, was elected a director and editor. Ray Latshaw, secretary/treasurer of MALA was named secretary/treasurer of the magazine and 510 out of 1,000 shares of stock were issued to MALA. Kermit Oswald was hired to assist Jones in operating the magazine. Six months later, the Board of Directors agreed that the success of *The Plan* was due to the past actions of Robert A. Jones and they agreed to transfer to him 260 shares of the stock of *The Plan*, Inc., of the 510 shared owned by the Association. Included with the minutes of the board was an agreement between Jones and the association allowing MALA to buy back the shares at a fair market price upon Jones' death. If he ever wanted to sell the

Photo courtesy of EBMDA

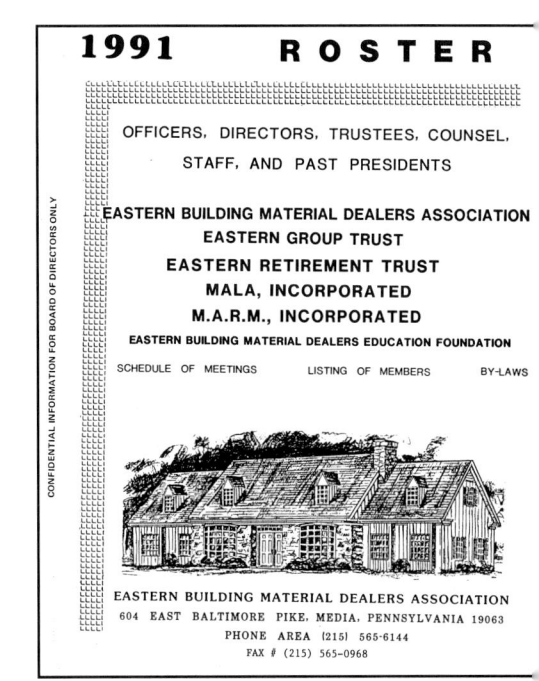

Photo courtesy of EBMDA

Photo courtesy of EBMDA

Photo courtesy of EBMDA

shares, the association had the right of first refusal.

It is apparent from the minutes, as well as from the remembrances of former presidents of the association, that his arrangement was a way of providing additional income to Jones at a time when the association could not afford to increase his salary. When Jones took a medical leave of absence in 1951, the association agreed to guarantee an income of $700 a month from *The Plan* for the duration of the leave.

Once again, after this reorganization, the magazine disappears from the minutes of the Board of Directors until November 1962, when there was an emergency meeting of the Executive Committee to deal with an anticipated deficit in the association's accounts due in large measure to declining revenues from dues. As part of the review of the operation of the association, it was reported that *The Plan* had an operating loss for the previous ten years. Yet another committee was appointed to look into the operation of the magazine in an effort to make it profitable again. By the end of the year, the Board of Directors agreed to cancel a $6,000 debt from *The Plan* as a bad debt. The committee reported that they were continuing their review of the operation of the magazine. By mid 1963, the special committee reported that they had rejected eliminating the publication because it was a necessary organ of the association and that they had agreed to work to make it more economically viable. After rejecting having MALA take over the magazine, granting an outright subsidy to *The Plan* by MALA, or having *The Plan* take over the publication of *The Dealers Directory and Buyers Guide,* the committee recommended a concerted effort to increase revenue by having members of the Board of Directors contact potential advertisers and cutting costs of production.

These were only stopgap measures. By 1969, *The Plan* had a printing debt of $8,000 and advertising revenues were declining. The board discussed four options: have Bob Jones spend more time getting ads; dissolve the magazine; sell it to some other organization; or have MALA purchase it. The board took no action at this meeting. However, the magazine suspended publication in 1969, and a proud sixty-year tradition and effective mouthpiece for the association creased to exist.

"Splinters," by Bill McEwing, was a popular column in The Plan *and was an easy way to keep track of the comings and goings of fellow lumber dealers. Photo courtesy of EBMDA*

140

SPLINTERS
LITTLE POINTED NEWS ITEMS PICKED UP HERE AND THERE

By BILL McEWING—1343 So. Lindenwood St., Philadelphia, Pa.

We enjoyed a nice letter from Charlie Sweigert of the Shillington Heights Building Yards, but was sorry to learn that Charlie after 21 years service as manager of this yard is retiring on April 15th. The owner died last fall and the trustees sold it out recently.

Charlie is a noted fisherman and he tells us to warn his friend Walt Johnston that Walt is not going to catch all the big rocks this year—sounds like Charlie is going to do some fishing and meditating. When the fish don't bite, we hope Charlie will enjoy this observation of Izaak Walton:

"No life, my honest scholar, no life so happy and so pleasant as the life of a well governed angler; for when the lawyer is swallowed up with business and the statesman is preventing or contriving plots, then we sit on cowslip banks, hear the birds sing and possess ourselves in as much quietness as these silent silver streams which we see glide so quietly by us. God never did make a more calm, quiet, innocent recreation than angling."

SPLINTERS

A billing clerk in the classified advertising office of the Tucson Daily Star wrote this on an overdue statement:

"We hope you found your dog; also we hope you pay this bill".

The payment came in, with this note:

"We hope you're glad we finally paid this and hope you're not mad at us for overlooking it so long. P.S. We found the dog, but we are sorry we did. She brought home four more. Your ads sure bring results. Do you want a little white pup?"

SPLINTERS

Bill McGarvey of Miller & McGarvey, Philadelphia, is one of the leaders in Volunteer Fire Association in New Jersey. Bill seems such a quiet, unassuming fellow that it is hard to believe he is a real honest to goodness fire-fighter but don't ring a bell near him. Charlie Miller says that every time the sirens go off at Saturday, 12:05 p. m. Bill grabs an ax and jumps right out of the office window.

SPLINTERS

Howard F. Scarborough, a well known lumber salesman, tells us that he and his good wife were down to Tampa, Florida, recently to see their son, Bob, before he left for overseas and that they stopped at Cummers cypress mills and several others. He tells us that Cummer has about 120 German war prisoners working at the mill and they sure looked as if they were well fed. Sort of gets a fellows goat says Howard to see them and then hear about the way our boys are treated by the Nazis in prison camps.

SPLINTERS

Bill Brooks of Joseph H. Sykes Lumber Co., Philadelphia, is getting spring fever again which means a trip down South. Bill might not be so fortunate in finding any mills that will condescend to take an order but we wager he will have a good southern meal that winds up with old fashioned Apple Jack.

SPLINTERS

God sews up the buds of flowers very tight and after a while He lets the sun and rain open the stitches.

SPLINTERS

Two rabbits sat watching a busy line on which thousands of tank parts were being made. Finally one observed: "I'm not jealous, but they must have started out with more than two.

SPLINTERS

A Cleveland G.I. stationed in California, was annoyed to receive an income tax notice indicating a tax payment was overdue. He sat down and wrote:

Roses are red,
Violets are blue,
I'm in the army,
To heck with you.

From the Bureau of Internal Revenue he received this unbureaucratic reply:

Your poem is fine,
And so are you,
So forget the tax
Till the war is through.

SPLINTERS

Jack Early of the Chestnut Hill Lumber Co., notices in the Real Estate News that McCloskey & McShain are two of the boys who are taking over a famous hotel in Atlantic City, and wonders if

---

"BEAR IN MIND"
B. C. BAER & SON, Reading, Pa.
White Pine, Yellow Pine, N. C. Pine, Hemlock, Spruce, Cypress, Hardwoods, West Coast Lumber and 24-inch B. C. Shingles
"DEPENDABLE STOCKS and SERVICE"
WHOLESALE LUMBER

W. S. ROHRBACH
WHOLESALE LUMBER
READING, PENNSYLVANIA
HIGH GRADE STOCKS ONLY

Time Line

Business and Economics

1890s
- Sherman Anti-Trust Act
- First can of pineapple produced
- Gillette Safety Razor invented
- Gold Rush begins

1900s
- U.S. Steel created by J. P. Morgan
- General Motors founded
- Panic of 1907
- Texas oil boom begins

1910s
- Income tax established
- F. W. Woolworth founded
- Federal Reserve System established
- First transcontinental telephone

1920s
- Ford introduces 40-hour work week
- Westinghouse opens first broadcast station, KDKA
- Dupont introduces cellophane
- Stock market crash of '29

1930s
- Empire State Building Opens
- Minimum wage of $.40 an hour
- Blue Eagle Campaign of NRA
- Social Security Act passed

1940s
- First electronic digital computer in Philadelphia
- Cortisone discovered
- Taft-Hartley Act passed over President Truman's veto
- National debt reaches $250 billion

1950s
- First Holiday Inn opens
- Link established between cancer and smoking
- Sputnik I launched
- AFL-CIO merges

1960s
- Japanese cars imported for first time
- Federal legislation for equal pay for males and females
- Medicare, financed by Social Security, proposed by President Kennedy
- Vietnam War

1970s
- Airlines deregulated
- Cigarette ads banned on TV by federal legislation
- Dow Jones closes above 1,000 for first time
- Bank prime rate goes to 14.5 percent

1980s
- Inflation reaches 12.4 percent
- "New Coke" flops as drinkers insist on "Classic Coke"
- AT&T divests Bell Telephone
- Worst stock crash in history—Dow Jones down 508 points

1990s
- "The Environmental Decade"
- National debt continues to grow above $2 trillion
- Communism collapses in Europe
- Capitalism emerges in Russia

Politics	Everyday Life
"Separate but equal" doctrine announced by Supreme Court Spanish-American War President Cleveland stops gold drain with loan from J. P. Morgan William McKinley elected president	Census—62,947,142 Zipper invented in Meadville, Pa. Modern Olympics founded Basketball invented
United States goes on the Gold Standard McKinley assassinated and Theodore Roosevelt becomes president Pure Food and Drug Act enacted William Howard Taft elected president	Census—75,994,575 Rotary Club founded First radio program of voice and music Wright Brothers flight
World War I Woodrow Wilson elected president for two terms Bolshevik Revolution Panama Canal opens	Census—91,972,266 Fathers' Day celebrated Bobbed hair for women sweeps nation Flu epidemic kills 22 million worldwide
Warren Harding, Calvin Coolidge and Herbert Hoover presidents of United States Nineteenth Amendment gives women right to vote Sacco and Vanzetti trial becomes cause celebre	Census—105,710,620 Prohibition goes into effect Television first demonstrated Ku Klux Klan resurgent in United States
Franklin Roosevelt elected to two terms as president Civilian Conservation Corps established FDR's attempt to "pack" Supreme Court fails Federal Reserve System reorganized	Census—122,755,046 Great Depression Prohibition repealed Al Capone indicted for income tax evasion
Roosevelt elected to third and fourth terms G.I. Bill of Rights passed Federal government "guarantees" full employment Harry Truman becomes president upon death of FDR	Census—131,669,275 World War II Paperback books introduced to conserve paper Movies gross $2 billion
Eisenhower becomes president first Republican president in twenty years Supreme Court outlaws segregation in public education Interstate Highway system established and funded by federal government Summit Meeting fails with U-2 flight discovery	Census—150,697,361 Color TV first broadcast Korean War Americans move to the "suburbs"
John F. Kennedy assassinated and Lyndon Johnson becomes president Martin Luther King and Robert F. Kennedy assassinated Berlin Wall goes up Cuban Missile Crisis	Census—179,243,500 Americans advised to build fallout shelters Payola scandals hit TV and radio Americans walk on the moon
OPEC Oil Embargo Watergate Scandal and President Nixon resigns rather than face impeachment Gerald Ford becomes president and loses reelection bid to Jimmy Carter Camp David Accord brings peace between Egypt and Israel	Census—205,000,000 Vietnam War ends Long lines at gas stations Three Mile Island nuclear accident
Ronald Reagan becomes president Reagan fires 13,000 striking air traffic controllers Graham Rudman Act enacted to try to fight federal deficit George Bush becomes president	Census—226,547,000 Challenger space shuttle explodes, killing seven crew members Compact discs replace records and tapes Personal computers become popular
Berlin Wall comes down, Germany reunited Iraq invades Kuwait and U.S. and United Nations allies send troops to force removal USSR breaks apart Growth of national debt continues to plague the United States	Census—251,000,000 Drugs become a major problem The AIDS epidemic continues Sexual harassment enters the American consciousness

Then and Now

(top photo) Atticks and Britchers, late 1880s.
(bottom photo) J. H. Rearick and Son, Inc., today.
Photos courtesy of J. H. Rearick and Son, Inc.

(top photo) Smick and Harris Lumber Company, 1907.
(bottom photo) Smick Lumber and Building Materials, today.
Photos courtesy of Smick Lumber and Building Materials Center

*(top photo) Frisbee Lumber Company around the turn of the century.
(bottom photo) Lehigh Lumber Company. The building was built in 1948 and burned down in 1987.
Photos courtesy of Lehigh Lumber Company*

(top photo) J. T. and L. E. Eliason, Inc., 1919.
(bottom photo) Brosius and Eliason today.
Photos courtesy of Brosius-Eliason, Inc.

*(top photo) Passmore Supply, circa 1930.
(bottom photo) Pyles Home and Supply today.
Photos courtesy of Pyle's Home and Supply*

148

(top photo) Robert E. Cook & Son, 1938.
(bottom photo) Elverson Supply Today.
Photos courtesy of Elverson Supply

Chief Staff Officers

During the course of its 100 years, the association has been fortunate to have chief staff officers with long tenures in office. From 1892 until 1902, the functions of the staff officer were done by the elected secretary of the association, who was a member and changed almost yearly. Beginning in 1902 that position was no longer elected by the membership with the other officers, but, rather, was appointed by the Board of Directors. From 1902 to 1909 the position of secretary was held by B. F. Laudig.

J. Frederick Martin
Secretary/Treasurer
(1910–1942)
Photo courtesy of EBMDA

Robert A. Jones
Manager and Executive Vice President
(1941–1970)
Photo courtesy of EBMDA

Russell J. Allen
Executive Vice President
(1970–1978)
Photo courtesy of EBMDA

*David B. Kreidler
Executive Vice President
(1978–present)
Photo courtesy of EBMDA*

Association Offices

*1910–1922
608 Bulletin Building
Philadelphia, Pa.
Photo courtesy of Free Library of Philadelphia*

*1922–1935
212 Otis Building Sixteenth and Sansom Streets
Philadelphia, Pa.
Photo courtesy of Free Library of Philadelphia*

*1935–1941
1102 Girard Trust Building
Philadelphia, Pa.
Photo courtesy of Mellon PSFS*

*1941–1954
Integrity Bank Building
1528 Walnut St.
Philadelphia, Pa.
Photo courtesy of Free Library
of Philadelphia*

*1959–1966
2 Penn Center Plaza
Philadelphia, Pa.
Photo courtesy of Free Library
of Philadelphia*

1966–1976
First Pennsylvania Bank Building
7 East Lancaster Ave.
Ardmore, Pa.
Photo courtesy of EBMDA

1954–1959
1523 Walnut St.
Philadelphia, Pa.
Photo courtesy of EBMDA

1976–present
604 E. Baltimore Pike
Media, Pa.
Photo courtesy of EBMDA

Presentation of Century Club Certificates
(top left) Harry H. Johnson III (left) presents certificate to Howard E. Heckler, president, Est. of George S. Snyder, Inc., Hatfield, Pa.
(top right) Left to right: James Clinger, president, Clinger Lumber Company, Milton, Pa., receives certificate from Harry H. Johnson III.
(bottom) Left to right, Lawrence Freas, Jr. and L. N. Freas of N. K. Freas & Sons, Andalusia, Pa., receives certificate from David B. Kreidler.
Photos courtesy of EBMDA

Centennial Companies

(Member Companies who have been in existence 100 years or more)

1785 Tinsman Brothers, Inc., Lumberville, Pa.
1845 Justice Lumber Co., Inc., Carney's Point, N.J.
1850 Walbert Lumber, Mertztown, Pa.
1852 Schmuck & Company, Hanover, Pa.
1854 Clinger Lumber Company, Milton, Pa.
1854 Watson Malone & Son, Haverford, Pa.
1857 Long and Bomberger, Lititz, Pa.
1860 Eberly Lumber Company, Mechanicsburg, Pa.
1861 E. S. Adkins & Co., Salisbury, Md.
1863 A. K. Shearer Company, North Wales, Pa.
1863 Smith & Reifsnider, Inc., Westminister, Md.
1863 William H. Fritz, Inc., Berwyn, Pa.
1868 Wm. D. Bowers Lumber Co., Woodsboro, Md.
1868 Raup Lumber & Construction, Shamokin, Pa.
1872 Newark Lumber Company, Newark, Del.
1873 Frost-Watson Lumber Corp., Newtown, Pa.
1874 Elverson Supply Company, Elverson, Pa.
1878 G. A. Miller Lumber Co., Inc., Williamsport, Md.
1878 J. C. Snavely & Son, Inc., Lancaster, Pa.
1879 Coyle Lumber Company, Carlisle, Pa.
1880 N. K. Freas Sons, Andalusia, Pa.
1881 Cavetown Planing Mill, Cavetown, Md.
1882 Brosius-Eliason Co., Newark, Del.
1883 Piper Building Supply, Oakland, Md.
1884 Estate of George S. Snyder, Inc., Hatfield, Pa.
1885 The Jacob Gehron Co., Williamsport, Pa.
1885 Kneas March Lumber Co., Norristown, Pa.
1886 I. F. March Sons, Bridgeport, Pa.
1888 J. E. Mitchell Co., Glenolden, Pa.
1889 Peoples Lumber & Supply Co., Inc., Mount Airy, Md.
1889 W. T. Galliher & Bro., Inc., Springfield, Va.
1890 Ritter & Smith, Allentown, Pa.
1890 J. P. Collins Company, Inc., South Seaville, N.J.
1890 H. M. Stauffer & Sons, Inc., Leola, Pa.
1890 Fessenden Hall, Inc., Pennsauken, N.J.
1890 Bangor Lumber Co., Bangor, Pa.
1891 J. H. Rearick & Son, Inc., Dillsburg, Pa.
1892 K & L Company, Inc., Quakertown, Pa.

Harry Blair, Jr.
C.A. Niece & Co., Inc.
Lambertville, NJ

Arthur R. Borden
Lewisburg Bldrs. Sply. Co.
Lewisburg, PA

Charles M. Bowers
The Wm.D.Bowers Lumber Co.
Woodsboro, MD

Aldo Braido
General Supply Company
Easton, PA

Perry E. Brunk
People's Supply Co., Inc.
Hyattsville, MD

Randall Brunk
People's Supply Co., Inc.
Hyattsville, MD

Robert M. Bushey
The Cavetown Planing Mill Co.
Cavetown, MD

Jon C. Clapper
Clapper's Bldg. Materials, Inc.
Meyersdale, PA

Edward H. Davis, Jr.
Moulton H. Davis Est., Inc.
West Chester, PA

Gene S. DiMedio
DuBell Lumber Company
Cedar Brook, NJ

John H. Eaton, Jr.
Barrons Enterprises, Inc.
Gaithersburg, MD

Lawrence D. Forman
H.B. Trueman Lumber Company
St. Leonard, MD

Farrell L. Goble
Brosius-Eliason Company
New Castle, DE

Roland L. Green, Jr.
Hyatt Bldg. Supply Co., Inc.
Damascus, MD

James P. Rauch
Chairman of the Board
Crafton Lumber & Supply Co.
Pittsburgh, PA

Bruce C. Ferretti
Vice Chairman/Treasurer
Lehigh Lumber Company
Bethlehem, PA

Gerald S. Greene
Christy's Supplies
Medford, NJ

Lee R. Harman
U.L. Harman, Inc.
Marydel, DE

Elected Leadership 1991

Eastern Building Material Dealers Association
604 E. Baltimore Pike
Media, PA 19063-1795
(215) 565-6144 • (412) 561-2323
FAX (215) 565-0968

1892 – 1992

William D. Hayes
Moxham Lumber Company
Johnstown, PA

W. James Hollenbach
W. Hollenbach Company
Boyertown, PA

Terry L. Kauffman
Reinholds Lmbr. & Mlwrk., Inc.
Reinholds, PA

James D. Kolker
Maryland Lumber Company
Baltimore, MD

W. Russell Lamar, Jr.
Lamar & Wallace, Inc.
Landover, MD

O. Grant Little
Little Lumber Company, Inc.
Benton, PA

Brian J. Lucas
Gilbert Lumber & Supply Co.
McKeesport, PA

G. Eugene Mackie
Mackie's Home Center
Cecilton, MD

J. Kirk Miller
Sutersville Lumber Co., Inc.
Sutersville, PA

Steven D. Mitchell
G.R. Mitchell, Inc.
Refton, PA

Van T. Mitchell
MSI, Inc.
LaPlata, MD

James D. Neal
The Nuttle Lumber Company
Denton, MD

G. Robert Overhiser
Collingdale Millwork Company
Collingdale, PA

Dale W. Parker
Cramer's Home Centers
E. Stroudsburg, PA

David G. Patterson
Patterson Lumber Company
Wellsboro, PA

Mary K. Rearick
J.H. Rearick & Son, Inc.
Dillsburg, PA

Jay F. Risser
J.H. Brubaker, Inc.
Lancaster, PA

J. Fred Robinson
Newark Lumber Company
Newark, DE

Edwin F. Scholtz
Sykes-Scholtz-Collins Lumb., Inc.
Philadelphia, PA

Gregory Shelly
Shelly Enterprises
Souderton, PA

R. Brian Shober
NELCO Lumber & Home Centers
Hamburg, PA

B. Harold Smick, Jr.
Smick Lmbr. & Bldg. Mat'ls. Ctr.
Quinton, NJ

John E. Smith, Jr.
Smith Bldg. Supply, Inc.
Churchton, MD

H. Paul Starr, Jr.
H.P. Starr & Sons, Inc.
Valencia, PA

Vincent J. Tague
Tague Lumber, Inc.
Philadelphia, PA

William K. Turner
E.S. Adkins & Company
Salisbury, MD

David Waitz
Emily Lumber Company
Philadelphia, PA

Karl J. Westover
Allensville Planing Mill, Inc.
Allensville, PA

Past Presidents

Year	President	Town	Firm
1892	James A. O'Reilly	President of Temporary Organization	
1892	F. P. Heller	Reading, Pa.	
1893–98	S. H. Sturdevant	Wilkes Barre, Pa.	Sturdevant & Graf
1899–1907	W. M. James	Steelton, Pa.	Steelton Planing Mill & Lumber
1908	T. J. Snodon	Scranton, Pa.	Mason & Snowdon
1909	S. C. Creasy	Bloomsburg, Pa.	Creasy & Wells
1910–11	Henry Palmer	Langhorne, Pa.	Henry Palmer Co.
1912	C. Frank Williamson	Media, Pa.	
1913–14	J. J. Milleisen	Mechanicsburg, Pa.	Milleisen & Son
1915	Theodore A. Mehl	Rosemont, Pa.	Mehl & Latta
1916	William S. Goff	Wilkes Barre, Pa.	Goff Lumber Company
1917	Albert J. Thompson	Wycombe, Pa.	Albert J. Thompson
1918	E. K. Moyer	Perkasie, Pa.	J. H. Moyer & Son
1919–20	Harry J. Meyers	Bethlehem, Pa.	Brown-Borhek Co.
1921–23	Fred H. Ludwig	Reading, Pa.	Merritt Lumber Yards
1924–25	Wilson H. Lear	Philadelphia, Pa.	Wilson H. Lear
1926–26	Luther C. Ogden	Cape May, N.J.	Geo. Ogden & Sons
1928	John H. Derr	Philadelphia, Pa.	John H. Derr Co.
1929–30	G. C. Rosser	Nanticoke, Pa.	Susquehanna Lumber Co.
1931–32	J. T. Eliason, Jr.	New Castle, Del.	J. T. & L. E. Eliason
1933–34	Bruce H. Helfrich	Baltimore, Md.	Geo. H. Helfrich & Son
1935–36	Dr. A. M. Northrup	Ashley, Pa.	Bowden Northrup Co.
1937	Horace B. Wilgus	Philadelphia, Pa.	Philadelphia Reserve Supply
1938–39	Joseph W. Brosious	Wilmington, Del.	Brosious & Smedley
1940	George P. Kingsley	Bethlehem, Pa.	Brown & Borhek Co.
1941–43	J. Hammond Geis	Baltimore, Md.	Jno H. Geis & Co.
1944–45	Elias W. Nuttle	Denton, Md.	Nuttle Lumber & Coal Co.
1946–48	Watson Malone III	Philadelphia, Pa.	Watson Malone Sons
1949–51	Claude G. Ryan	Lancaster, Pa.	Jno D. Bogar Lumber Co.
1952–53	G. Hunter Bowers	Frederick, Md.	Wm. D. Bowers Lumber Co.
1954	W. R. Lamar	Washington, D.C.	Lamar & Wallace
1955	Hugh M. Peter	Pleasantville, N.J.	Peter Lumber Co.
1956	L. H. Schmoyer	Boyertown, Pa.	L. H. Schmoyer
1957	Frank S. Buechley	Hamburg, Pa.	Buechley Millwork & Lumber
1958	Frank M. Hankins, Jr.	Bridgeton, N.J.	H. H. Hankins & Bro.
1959	Charles D. Hummer	Chester, Pa.	Hummer & Green
1960	James C. Dillon	Wilmington, Del.	Wilmington Sash & Door Co.
1961–62	John W. Lundy	Williamsport, Pa.	Lundy Lumber Company
1963	William T. Wolf	York, Pa.	Wolf Supply Company
1964	S. F. M. Adkins, Jr.	Easton, Md.	E. S. Adkins & Company
1965	Jesse Snavely, Jr.	Landisville, Pa.	J. C. Snavely & Sons, Inc.
1966–67	William E. Norman	Gaithersburg, Md.	Gaithersburg Lumber & Supply Co.
1968	James T. Eliason III	New Castle, Del.	Brosius-Eliason
1969–70	Joseph E. Haenn, Jr.	Frazer, Pa.	Jos. E. Haenn, Inc.

1970–Title of elected dealer head of organization changed from President to Chairman of the Board

Year	President	Town	Firm
1971	Clifford L. Cramer	East Stroudsburg, Pa.	Cramer's Cashway
1972–73	G. Hunter Bowers, Jr.	Frederick Md.	Wm. D. Bowers Lumber Co.
1974–75	B. Harper Beatty, Jr.	Upper Darby, Pa.	Beatty Lumber & Millwork Co.
1976–77	Edgar B. Harman	Marydel, Del.	U. L. Harman, Inc.
1978–79	John D. Mitchell	LaPlata, Md.	Mitchell Supply Co.
1980–81	Richard A. Kauffman	Lemoyne, Pa.	Otto & Hollinger, Inc.
1982–83	William S. Smith, Sr.	Gaithersburg, Md.	Gaithersburg Lumber & Supply Co.
1984–85	J. Bill Kildoo	New Castle, Pa.	Kildoo & Cooper Inc.
1986–87	B. Harold Smick, Jr.	Quinton, N.J.	Smick Lumber & Building Material Center
1988–89	Al Braido	Easton, Pa.	General Supply Company
1990–91	James P. Rauch	Pittsburgh, Pa.	Crafton Lumber & Supply Co.

1991 Gavel Cavaliers

(Living Past Chief Elected Officers)

Eastern Building Material Dealers Association
604 E. Baltimore Pike • Media, PA 19063-1795
(215) 565-6144 • (412) 561-2323 • FAX (215) 565-0968

G. Hunter Bowers, Sr.
1952-1953
The Wm.D. Bowers Lmbr. Co.
Woodsboro, MD

Hugh M. Peter
1955
Peter Lumber Co.
Pleasantville, NJ

Frank M. Hankins, Jr.
1958
H.H. Hankins & Bro.
Bridgeton, NJ

John W. Lundy
1961-1962
R&J Realty Company
Williamsport, PA

William T. Wolf
1963
Wolf Supply Company
York, PA

William E. Norman
1966-1967
Barrons Gaithersburg Lumber
Gaithersburg, MD

James T. Eliason, III
1968
Brosius-Eliason Company
New Castle, DE

Joseph E. Haenn, Jr.
1969-1970
Joseph E. Haenn, Inc.
St. David, PA

Clifford L. Cramer
1971
Cramer's Home Center
E. Stroudsburg, PA

G. Hunter Bowers, Jr.
1972-1973
The Wm. D. Bowers Lmbr. Co.
Woodsboro, MD

B. Harper Beatty, Jr.
1974-1975
Beatty Lumber & Millwork Co.
Upper Darby, PA

Edgar B. Harman
1976-1977
U.L. Harman, Inc.
Marydel, DE

Richard A. Kauffman
1980-1981
Otto & Hollinger, Inc.
Lemoyne, PA

William W. Smith, Sr.
1982-1983
Barrons Gaithersburg Lumber
Gaitherburg, MD

J. Bill Kildoo
1984-1985
Kildoo & Copper, Inc.
New Castle, PA

B. Harold Smick, Jr.
1986-1987
Smick Lmbr. Bldg. Mat'l. Ctr.
Quinton, NJ

Aldo Braido
1988-1989
General Supply Company
Easton, PA

James P. Rauch
1990-1991
Crafton Lumber & Supply
Pittsburgh, PA

Active Dealer Members

(As of January 1991)
Eastern Building Material Dealers Association

Company Name	City, State
A & A Lumber Supply Company	Avondale, Pa.
A & S Building Supply, Inc.	Doylestown, Pa.
Ace Lumber & Millwork Co., Inc.	Philadelphia, Pa.
Addlesberger Building Supplies, Inc.	Chambersburg, Pa.
*E. S. Adkins and Company	Easton, Md.
*E. S. Adkins and Company	Chestertown, Md.
E. S. Adkins and Company	Salisbury, Md.
*The Adkins Co. of Ocean City	Ocean City, Md.
*The Adkins Company	Berlin, Md.
The Adkins Company	Berlin, Md.
*E. S. Adkins and Company	Seaford, Del.
Allegheny Lumber & Supply Co.	Tarentum, Pa.
*Allensville Planing Mill	Lewistown, Pa.
Allensville Planing Mill	Allensville, Pa.
*Allensville Planing Mill	New Enterprise, Pa.
Allentown-Bethlehem Lumber Co.	Allentown, Pa.
Allibone Brothers	Stratford, N.J.
Allied Building Center, Inc.	Salisbury, Md.
Allied Building Products Corp.	East Rutherford, N.J.
Allstate Building Supply Co., Inc.	Lutherville, Md.
American Lumber Corporation	Baltimore, Md.
American Building Products Inc.	Jessup, Md.
APM, Inc.	Arendtsville, Pa.
Arey Lumber Company, Inc.	Wysox, Pa.
Arnold Lumber & Supply Company	Red Lion, Pa.
Arnold Lumber Company	New Kensington, Pa.
Artistic Furnishings, Inc.	Trumbauersville, Pa.
Associated Lumber & Mfg. Co., Inc.	Braddock, Pa.
Athens Home Center	Athens, Pa.
John H. Auld & Brothers Co.	Allison Park, Pa.
B & G Lumber Company	Elizabethtown, Pa.
Bangor Lumber Company	Bangor, Pa.
Bangor Lumber Company	Nazareth, Pa.
Barker Lumber Company, Inc.	Washington, D.C.
Barrons Enterprises, Inc.	Gaithersburg, Md.
Fred E. Beachy Lumber Company	Oakland, Md.
Norman A. Bean & Son, Inc.	Royersford, Pa.
Beatty Lumber & Millwork Co.	Upper Darby, Pa.
Robert L. Benjamin Co., Inc.	North East, Md.
Ira Berger & Sons, Inc.	Freeland, Pa.
*Berks Products	Temple, Pa.
*Berks Products	Kutztown, Pa.
Berks Products	Reading, Pa.
Berlin Lumber Company	Berlin, Pa.
Beyer Lumber Company	Riverside, Pa.
Big Value Builders, Inc.	Franklin, Pa.
*Delta Lumber	Perryville, Md.
Bishop Wood Products, Inc.	Elroy, Pa.
Jno D. Bogar & Son Company	Steelton, Pa.
Bond Home Center	Frostburg, Md.
E. W. Bostwick, Inc.	Elmer, N.J.
Wm. D. Bowers Lumber Company	Woodsboro, Md.
Bowmar Millwork Corporation	Mardela Springs, Md.
Bristol Fuel & Building Supply	Bristol, Pa.
Bristol Millwork, Inc.	Philadelphia, Pa.
Brookside Lumber & Supply Co., Inc.	Bethel Park, Pa.
Brookville Lumber Company	Brookville, Pa.
Brosius-Eliason Company	New Castle, Del.
*Brosius-Eliason Company	Wilmington, Del.
*Brosius-Eliason Company	Glasgow, Del.
J. H. Brubaker, Inc.	Lancaster, Pa.
*J. H. Brubaker, Inc.	Manheim, Pa.
Bryans Road Bldg. & Supply Co., Inc.	Bryans Road, Md.
Buchanan Lumber Company	Cumberland, Md.
Buffalo Lumber & Hardware, Inc.	Lewisburg, Pa.
John S. Buffenmyer Company	Mt. Joy, Pa.
Build-All Supply Company	Philadelphia, Pa.
Builder's Supply & Lumber Co., Inc.	Frederick, Md.
Buiders Choice	Elkton, Md.
Builders Prime Window	Bridgeport, Pa.
*Butz Building Center	Brodheadsville, Pa.
*Butz Building Center	Blakeslee, Pa.
*Butz Building Center	Lehighton, Pa.
Butz Building Center	Palmerton, Pa.
Buy-Rite Home & Bldg. Supply Center	Pennsauken, N.J.
Buyers Marketing Service, Inc.	Salisbury, Md.
C & C Builders Supply Co., Inc	Tarrs, Pa.
C D Millwork Co., Inc.	Collegeville, Pa.
Calvert Lumber Company, Inc.	Sharon, Pa.
Canter Lumber, Inc.	Leechburg, Pa.
Cape May Lumber Co., Inc.	Cape May, N.J.
Casselman Lumber	Grantsville, Md.
Castles Lumber Company	Carlisle, Pa.
Cavetown Planing Mill Company	Cavetown, Md.
Centennial Millwk. & Bldg. Supply Co.	Allentown, Pa.
Champion Lumber Company, Inc.	Champion, Pa.
*Chapin Lumber & Supply Co–Realty	Kingston, Pa.
Chapin Lumber & Supply Company	Temple, Pa.
Cherry Hill Builders Supply	Maple Shade, N.J.
Chopp & Company Incorporated	Waldorf, Md.
*Christy's Supplies, Inc.	Berlin, N.J.
Christy's Supplies, Inc.	Medford, N.J.
Church Hill Lumber Company	Church Hill, Md.
Clappers Building Materials, Inc.	Meyersdale, Pa.
Clarion Builders Supply Co.	Clarion, Pa.
Clasters	Bellefonte, Pa.
*Clasters	Altoona, Pa.
*Clasters	Bloomsburg, Pa.
*Clasters	DuBois, Pa.
*Clasters	Lewistown, Pa.
*Clasters	Lock Haven, Pa.
*Clasters	Milton, Pa.
*Clasters	Phillipsburg, Pa.
*Clasters	State College, Pa.
*Clasters	Sunbury, Pa.
*Clasters	Huntungdon, Pa.
Clay-Wood Lumber Company	Kittanning, Pa.
Clements & Steed, Inc.	Mt. Airy, Md.
Clinch-Tite Corporation	Sandy Lake, Pa.
Clinger Lumber Company	Milton, Pa.
O. C. Cluss Lumber Company	Uniontown, Pa.
*O C. Cluss Lumber Company	Greensburg, Pa.
Clymer Builders Supply, Inc.	Clymer, Pa.
Cobilt Incorporated	Martinsburg, Pa.

H. L. Coffman Lumber Company	Hagerstown, Md.	Dunmore Lumber Company	Dunmore, Pa.
*Collingdale Home Supply	Collingdale, Pa.	Dyson Lumber Company, Inc.	Great Mills, Md.
Collingdale Millwork Company	Collingdale, Pa.	East End Lumber Company	Williamsport, Pa.
*Collingdale Millwork Company	Frazer, Pa.	Eberly Lumber Company	Mechanicsburg, Pa.
*Collingdale Millwork Company	Newark, Del.	Eckman Lumber Company	Lehighton, Pa.
Collins Supply Company, Inc.	Wilmington, Del.	Eisenhardt Mills, Inc.	Easton, Pa.
J. P. Collins Company	South Seaville, N.J.	Les Elliot Company	Smithburg, Md.
Colson's Home & Building Center	North Wildwood, N.J.	Elverson Supply Company, Inc.	Elverson, Pa.
*Colson's Home & Building Center	Rio Grande, N.J.	Emily Lumber Company	Philadelphia, Pa.
Colucci Lumber Company	Malaga, N.J.	Erco Ceilings, Inc.	Glassboro, N.J.
Concord Plywood & Supply Co.	Upland, Pa.	Erdman Lumber Company	Baltimore, Md.
Contract Door & Hardware, Inc.	Wilmington, Del.	Erwine's, Inc.	Bloomsburg, Pa.
Conyngham Builders & Supply Co.	Conyngham, Pa.	Evergreen Lumber Company, Inc.	Woodbury, N.J.
Cook Brothers Supply Company	Darlington, Pa.	Fagen's, Inc.	Verona, Pa.
Copper Lumber, Inc.	New Castle, Pa.	Fahringer Distributors, Inc.	Berwick, Pa.
Cottman/Stoneback, Inc.	Bristol, Pa.	*Fairmount Mill & Lumber Co.	Baltimore, Md.
Coyle Lumber Company	Carlisle, Pa.	Faxon Lumber Company	Williamsport, Pa.
Crafton Lumber & Supply Co.	Pittsburgh, Pa.	J. Fazzio, Inc.	Glassboro, N.J.
Crain Lumber Company	Port Matilda, Pa.	Leland L. Fisher, Inc.	Rockville, Md.
Cramer's Cashway, Inc.	East Stroudsburg, Pa.	522 Lumber & Home Center	Kreamer, Pa.
Creegers Home & Hardware, Inc.	Rising Sun, Md.	Flower Lumber Company	Sharon, Pa.
Croft Lumber Company, Inc.	Sayre, Pa.	Fluder Home & Building Supply, Inc.	Windber, Pa.
V. B. Cross Lumber Company	High Falls, N.Y.	Ford Lumber Company, Inc.	Upper Marlboro, Md.
Cummings Lumber Company, Inc.	Troy, Pa.	*Ford Lumber Company	Fort Washington, Md.
Currie Lumber & Millwork Co.	Philadelphia, Pa.	Forge Millwork & Lumber Co.	Malvern, Pa.
D "n" J Lumber & Millwork Co.	Philadelphia, Pa.	Forty-Fort Lumber Company, Inc.	Forty-Fort, Pa.
Dale Lumber Company, Inc.	Falls Church, Va.	Fox Chapel Stone & Supply, Inc.	Pittsburgh, Pa.
Dambach Lumber & Supply Co.	Harmony, Pa.	Franks Lumber Company, Inc.	Wilkes-Barre, Pa.
Danby Lumber Company	Chadds Ford, Pa.	N. K. Freas' Sons	Andalusia, Pa.
Darby Lumber Company	Darby, Pa.	Freehling Lumber Co., Inc.	Cabot, Pa.
Davey Lumber Company	Dover, Del.	French Lumber Company	Mt. Holly, N.J.
Moulton H. Davis Lumber	West Chester, Pa.	Frey Lumber Company	Smithfield, Pa.
*Moulton H. Davis Lumber	Eagle, Pa.	Friel Lumber Company	Queenstown, Md.
Decker Lumber Company	Tunkhahnock, Pa.	William H. Fritz, Inc.	Berwyn, Pa.
Deepwater Lumber & Supply Co., Inc.	Pennsville, N.J.	Frost-Watson Lumber Corp.	Newtown, Pa.
DeGrange Lumber Center Glen	Burnie, Md.	G & W Lumber, Inc.	Accident, Md.
Delaware Cedar Company, Inc.	Lewes, Del.	G. C. Corporation	Williamsport, Pa.
*Delaware County Supply Company	Glen Mills, Pa.	Galliher & Hugely Assocs., Inc.	Washington, D.C.
Delaware Lumber & Millwork	Dover, Del.	W. T. Galliher & Brothers, Inc.	Springfield, Va.
Delmont Builders Supply, Inc.	Delmont, Pa.	Gemco Home Improvement Products	Tuchannock, Pa.
Delta Lumber Do-It Center	Whiteford, Md.	General Supply Company, Inc.	Easton, Pa.
Denlinger, Inc.	Paradise, Pa.	Wm. A. Geppert, Inc.	Roslyn, Pa.
Denver Planning Mill	Denver, Pa.	Gerhart Brothers	Ephrata, Pa.
Desmet Lumber and Supply Co., Inc.	Cecil, Pa.	Gilbert Lumber & Supply Co.	McKeesport, Pa.
Devlin Lumber & Supply Co.	Rockville, Md.	*Glenside Lumber & Coal Company	Glenside, Pa.
Devon Building Supply Company	Devon, Pa.	Charles H. Goebel & Sons, Inc.	Folsom, Pa.
Donghia Lumber Company	Vandergrift, Pa.	Gosnell Bros. Lumber & Bldg. Supply	Altoona, Pa.
Donora Lumber Company	Donora, Pa.	Graeber's Lumber Company	Fairless Hills, Pa.
Dover Millwork, Inc.	Harrington, Del.	Great Northeastern Lmbr. & Millwork	Philadelphia, Pa.
*Doylestown Lumber & Millwork Co.	Doylestown, Pa.	H. G. Green Lumber Company	Mt. Royal, N.J.
Dravis Lumber Company, Inc.	Johnstown, Pa.	Grego Lumber Company	Bradford, Pa.
Dries Do-It Center	Macungie, Pa.	Grover Lumber Company	Princeton, N.J.
Du Bell Lumber Company	Cedar Brook, N.J.	Grubb Lumber Company, Inc.	Wilmington, Del.
*Du Bell Lumber Company	Cherry Hill, N.J.	Gumble Brothers, Inc.	Paupack, Pa.
Dubin Brothers Lumber Co., Inc.	Philadelphia Pa.	Haddon Fence Company	Mt. Holly, N.J.
Dukes Lumber Company, Inc.	Laurel, Del.	Haddonfield Lumber Company	Haddonfield, N.J.
*Dukes Lumber Company, Inc.	Seaford, Del.	Hagerstown Lumber Company	Hagerstown, Md.
Dundalk Lumber Company, Inc.	Baltimore, Md.	George M. Hall, Inc.	Pittsburgh, Pa.
*Dunkirk Supply Company	Dunkirk, Md	Hall's Home & Lumber, Inc.	Tioga, Pa.
Dunkirk Supply Company	Owings, Md.	The Hammer & Nail, Inc.	Equinunk Pa.

Hammond Builders Supply, Inc.	Corry, Pa.	Fred K. Kleinbach Sons, Inc.	Green Lane, Pa.
H. H. Hankins & Brothers	Bridgeton, N.J.	Klinedinst & Reichart, Inc.	Hanover, Pa.
Happel Lumber & Supply Company	Birdsboro, Pa.	Klinger Lumber Company, Inc.	Elizabethville, Pa.
Hardesty's True Value, Inc.	Springs, Pa.	H. H. Knoebel Sons, Inc.	Elysburg, Pa.
U. L. Harman, Inc.	Marydel, Del.	Kopp's Company, Inc.	Lineboro, Md.
*U. L. Harman, Inc.	Dover, Del.	Krupp, Meyers & Hoffman	Lansdale, Pa.
*U. L. Harman, Inc.	Hurlock, Md.	C. R. Dampman Lumber, Inc.	Pennsburg, Pa.
Harmon Lumber & Supply Company	McKeesport, Pa.	La Plata Mill & Supply Co., Inc.	La Plata, Md.
Harvey Lumber & Supply	Tarentum, Pa.	Lamigination	Telford, Pa.
Hatboro Lumber & Fuel Company	Hatboro, Pa.	Lamar & Wallace, Inc.	Landover, Md.
Hatfield Public Warehouse, Inc.	Hatfield, Pa.	Lapinski Lumber Company, Inc.	Mt. Carmel, Pa.
J. J. Hauser Lmbr. & Builders Supply	Greensburg, Pa.	Larry's Lumber & Supply, Inc.	Bloomsburg, Pa.
Hazel Builders, Inc.	Hazelton, Pa.	Laurel Building Supply Co., Inc.	Laurel Md.
Hecktown Building Supply	Bethelem, Pa.	Laurich Building Supply	Chambersburg, Pa.
*Heintz Home Centers Chadds	Ford, Pa.	Lee Lumberjack	Baltimore, Md.
H. G. Heintz, Inc.	Chester, Pa.	*Lehigh Lumber	Stroudsburg, Pa.
Hi-Way Supply Company	Dunbar, Pa.	*Lehigh Lumber	Pocono Lake, Pa.
High Point Building Products, Inc.	Tobyhanna, Pa.	Lehigh Lumber Company	Bethlehem, Pa.
Hitchner Lumber	Woodtown, N.J.	*Lehigh Lumber Singmaster Division	Macungie, Pa.
Hixon Hardware, Inc	Hancock, Md.	Lewisburg Builders Supply Co.	Lewisburg, Pa.
Hoffman Lumber Company	West Chester, Pa.	Lezzer Lumber Company	Curwensville Pa.
W. Hollenbach Company	Boyertown, Pa.	Liberty Building Supply Millwork	Philadelphia,, Pa.
E. J. Hollingsworth Company	Wilmington, Del.	Little Lumber Company	Benton, Pa.
Holt Lumber Company	Carbondale, Pa.	C. C. Lobb, Inc.	Havertown, Pa.
Home-Handyman Center, Inc.	Pleasantville, N.J.	Long & Bomberger	Lititz, Pa.
Homebuilders Warehouse	Lansdal,e Pa.	Charles C. Loose & Son, Inc.	Myerstown, Pa.
James DeSantis & Sons	Tuckahoe, N.J.	Al Lorenzi Lumber Company	Washington, Pa.
J. H. Hommer Lumber Co., Inc.	Glasgow, Pa.	*Lorenzi Lumber Company	McMurra,y Pa.
House of Paneling, Inc.	Pittsburgh, Pa.	Lucas Lumber	Levittown, Pa.
House of Paneling, Inc.	Wexford, Pa.	Lug & Tug True Value	Seneca, Pa.
O. W. Houts & Sons, Inc.	State College, Pa.	Lumadues, Inc.	Burlington, N.J.
Howard Lumber & Millwork, Inc.	Runnemede, N.J.	The Lumber & Millwork Company	Philadelphia, Pa.
Wm. Howeth, Jr., Inc.	Wittman, Md.	Lumber Products Company	Bristol, Pa.
Hyatt Building Supply Co., Inc.	Damascus, Md.	Luzerne Lumber Company	Luzerne, Pa.
Independent Lumber & Plywood Co.	Philadelphia, Pa.	Mace-Fremont Building Materials, Inc.	Baltimore, Md.
Inner Harbor Lumber & Hardware	Baltimore, Md.	Mackie's Home Center, Inc.	Cecilton, Md.
J & J Home Improvement Center	Elkton, Md.	Ernest Maier, Inc.	Bladensburg, Md.
J & L Building Materials, Inc.	Frazer, Pa.	Mansion Supply Company	Mt. Royal, N.J.
J. F. Mill & Lumber Company	New Castle, Pa.	Maryland Prime-O-Sash	Baltimore, Md.
Jarrettsville Hardware & Supply	Jarrettsville, Md.	I. F. March Sons, Inc.	Bridgeport, Pa.
Jem Lumber Company, Inc.	Bridgeville, Del.	*March Kneas Lumber Company	Norristown, Pa.
John's Sawmill & Lumber Yard	New Milford, Pa.	Marriotti Building Products	Old Forge, Pa.
Judge Lumber & Supply Company	Scranton, Pa.	C H Marshall, Inc.	Medi,a Pa.
Justice Lumber Company, Inc.	Carney's Point, N.J.	Marvic Supply Company, Inc.	Doylestown, Pa.
K & L Company, Inc.	Quakertown, Pa.	*Maryland Wood Preserving Corp.	Rockvill,e Md.
Clinton M. Kandle, Inc.	Pitman, N.J.	The Maryland Lumber Company	Baltimore, Md.
Kane Lumber, Inc.	Tamaqua, Pa.	*Masten Lumber & Supply Company	Millsboro, Del.
W. J. Kane, Jr., Inc.	Philadelphia, Pa.	Masten Lumber & Supply Company	Milford, Del.
KBC, Inc.	Honesdale, Pa.	*Masten Lumber & Supply Company	Clayton, Del.
Keehn Lumber & Supply, Inc.	Oley, Pa.	Maugansville Elevator & Lumber Co.	Maugansville, Md.
Kefauver Lumber Company	Forest Hill, Md.	Mayer, Inc.	Newark, Del.
*Kefauver True-Value Hardware	Belair, Md.	Mayerfeld Building & Plumbing Supply	Norma, N.J.
Kelly's Windows & Doors, Inc.	New Castle, Del.	Mays Landing Lumber & Bldg. Supply	MaysLanding, N.J.
L. M. Kennedy & Sons Company	Philadelphia, Pa.	McBroom's Hardware, Inc.	Latrobe, Pa.
Keystone Lumber & Truss Co., Inc.	Biglersville, Pa.	McCune Lumber Company	Shippensburg, Pa.
Keystone State Products Co., Inc.	Philadelphia, Pa.	McNany Lumber Company	Emlenton, Pa.
Fred Kiene Inc.	Philadelphia, Pa.	Mechanicsville Building Supply, Inc.	Mechanicsville, Md.
King Home Ctr. & Lumber Supply Co.	King of Prussia, Pa.	V. Menghini & Sons, Inc.	Hazelton, Pa.
J. F. Kingston Lumber Co., Inc.	Tafford, Pa.	Mercersburg Builders Supply Co., Inc.	Mercersburg, Pa.
Kirby's Paneling Center	Mechanicsburg, Pa.	Meredith-Roane Co., Inc.	Annapolis, Md.

Merritt Lumber & Home Center	Reading, Pa.	*Pace Supply Corporation	Pottstown, Pa.
Mid-State Building Supply, Inc.	Smithsburg, Md.	Panel-It Discount Center	Bridgeton, N.J.
Middlesex Building & Roofing Supply	Millsboro, Del.	The Paneling House, Inc.	New Castle, Pa.
Middletown Lumber, Inc.	Middletown, Pa.	Panther Valley Lumber	Lansford, Pa.
Midway Lumber Company, Inc.	Baltimore, Md.	Patterson Lumber Company, Inc.	Wellsboro, Pa.
Miles Building Supply Company	Pocomoke City, Md.	Peachy Builders	Belleville, Pa.
Robert G. Miller, Inc.	Hanover, Pa.	Peoples Coal & Supply Company	Stockerton, Pa.
G. A. Miller Lumber Co., Inc.	Williamsport, Md.	Peoples Supply Company, Inc.	Hyattsville, Md.
Mills Lumber & Supply Co.	Ft. Washington, Md.	The Peoples Lumber & Supply Co.	Mt. Airy, Md.
Millsboro Lumber & Millwork	Millsboro, Del.	Thomas W. Perry, Inc.	Chevy Chase, Md.
G. R. Mitchell, Inc.	Refton, Pa.	*Thomas W. Perry, Inc.	Gaithersburgh, Md.
J. E. Mitchell Company	Glenolden, Pa.	*Peter Lumber Company	Ocean City, N.J.
Mitchell Lumber & Fuel Oil	Pittston, Pa.	*Peter Lumber Company	Kennett Square, Pa.
*MSI, Inc.	Hughesville, Md.	*Peter Lumber Company	Limerick, Pa.
Mizell Lumber & Hardware, Inc.	Kensington, Md.	*Peter Lumber Company	Philadelphia, Pa.
Modern Cabinet & Construction Co.	Altoona, Pa.	*Peter Lumber Company	Egg Harbor City, N.J.
Montgomery Truss & Panel, Inc.	Grove City, Pa.	*Peter Lumber Company	Cornwells Heights, Pa.
P. M. Moore Company	Aliquippa, Pa.	*Peter Lumber Company	Columbia, Pa.
The Moxham Lumber Company	Johnstown, Pa.	*Peter Lumber Company	Pitman, N.J.
Moyer Lumber & Hardware	Bethlehem, Pa.	*Peter Lumber Company	Millville, N.J.
Amandus D. Moyer Lmbr. & Hardware	Gilbertsville, Pa.	*Peter Lumber Company	Medford, N.J.
*Amandus D. Moyer Lmbr. & Hrdwre.	Pottstown, Pa.	*Peter Lumber Company	Hammonton, N.J.
MRD Lumber Company, Inc.	Coopersburg, Pa.	Peter Lumber Company	Pleasantville, N.J.
MSI, Inc.	La Plata, Md.	*Peter Lumber Company	Collingswood, N.J.
Musselman Lumber, Inc.	New Holland, Pa.	Phillips Lumber Company	Stroudsburg, Pa.
*John H. Myers & Son, Inc.	Dallastown, Pa.	Phoenixville Lumber Company	Phoenixville, Pa.
*John H. Myers & Son, Inc.	York, Pa.	Piper Building Supply	Oakland, Md.
John H. Myers & Son, Inc.	York, Pa.	PJ's Home Center, Inc.	Peckville Pa.
Myers Lumber Company, Inc.	Hazelton, Pa.	Theo. B. Price	Cresco, ,Pa.
Narrowsburg Lumber Company	Narrowsburg, N.Y.	Pyles Home & Supply Center	Avondale, Pa.
Nelco Lumber & Home Centers	Hamburg, Pa.	*Quality Lumber & Bldg. Supply Co.	Oakford, Pa.
New Holland Custom Woodwork	New Holland, Pa.	"R" Lumber Center	Edgewater, Md.
New Jersey Hardware & Supply	Maple Shade, N.J.	Paul D. Rankin Builders Supplies	Clintonville, Pa.
Newark Lumber Company	Newark, Del.	Raup Lumber & Construction Co.	Shamokin, Pa.
Newtown Square Home Supply Co.	Newtown Square, Pa.	Reabold Coal & Building Supply	JimThorpe, Pa.
NFL Home Center & Building Supply	Shrewsbury, Pa.	J. H. Rearick & Son	Dillsburg, Pa.
Nicholson Lumber Company	Nicholson, Pa.	Reinholds Lumber & Millwork	Reinhold,s Pa.
C. A. Niece Co., Inc.	Lambertville, N.J.	Reistertown Lumber Company	Reistertown, Md.
L. G. Niles Lumber Co., Inc.	Wellsboro, Pa.	Reliable Lumber & Supply Co.	New Castle, Pa.
Nino's Home Care Center, Inc.	Philadelphia, Pa.	Rennekamp Lumber Company	Pittsburgh, Pa.
North Wales Millwork, Inc.	North Wales, Pa.	Rider's Lumber Company	Philadelphia, Pa.
Northampton Lumber Company	Northampton, Pa.	Ridge Lumber Company	Baltimore, Md.
Northern Building Products	Moscow, Pa.	Ridge Pike Building Materials, Inc.	Conshohocken, Pa.
*The Nuttle Lumber Company	Chester, Md.	J. T. Riley, Inc.	Philadelphia, Pa.
The Nuttle Lumber Company	Denton, Md.	The Ritchie Lumber Company	Baltimore, Md.
*The Nuttle Lumber Company	Rehobeth Beach, Del.	Rittenhouse Lumber & Millwork Co.	Erdenheim, Pa.
*The Nuttle Lumber Company	Millville, Del.	Ritter & Smith Company	Allentown, Pa.
O'Brien & Sons Brown	Mills, N.J.	Rock Hill Supply Company	Bala Cynwyd, Pa.
P. T. O'Malley Lumber Co., Inc.	Baltimore, Md.	Rockhall Lumber Company	Rock Hall, Md.
O'Neal Brothers, Inc.	Laurel, Del.	Roland's Special Millwork, Inc.	Chalfont, Pa.
Oakes & McClelland Company	Greenville, Pa.	Rotter's Home Center, Inc.	Lower Burrell, Pa.
Oakhill Enterprise, Inc.	Troy, Pa.	Rotz Lumber Yard	Chambersburg, Pa.
Ohio Valley Lumber Company	Ambridge, Pa.	Roxborough Building Supply	Philadelphia, Pa.
*On Center, Inc.	Athens, Pa.	Russell Brothers	Washington, Pa.
Oneida Lumber & Ace Hardware	Warren, Pa.	*Rust-Weir Supply Company	Georgetown, Del.
B. W. Otterman Lumber Company	Irwin, Pa.	Rybas Building Materials, Inc.	Philadelphia, Pa.
Otto & Hollinger, Inc.	Lemoyne, Pa.	*S & F Manufacturing Co., Inc.	Telford, Pa.
The Oxford Grain & Hay Co., Inc.	Oxford, Pa.	*S & S Lumber Company, Inc.	Pocono Lake, Pa.
Pace Supply Corporation	Lansdale, Pa.	S & S Lumber Company, Inc.	Scotrun, Pa.
*Pace Supply Corporation	West Chester, Pa.	*S & S Lumber Company, Inc.	Weissport, Pa.

Saco Supply, Inc	Timonium, Md.	Tinsman Brothers, Inc.	Lumberville, Pa.
Salisbury Builders' Supply, Inc.	Salisbury, Pa.	Tot'um Lumber & Supply Company	Pittsburgh, Pa.
Sandy Ponds Hardwoods	Quarryville, Pa.	Tri-City Lumber & Building Supplies	Vineland, N.J.
Saul Lumber Company	Apollo, Pa.	Triangle Bldg. Supplies & Lumber Co.	Quakertown, Pa.
Saylorsburg Lumber Company	Saylorsburg, Pa.	Triangle Building Supply, Inc.	Bellefonte, Pa.
The Schaeffer Lumber Company	Westminster, Md.	Trim-Rol Building Products, Inc.	Greensburg, Pa.
Schmoyer's Lumber Company	Schwenksville, Pa.	H. B. Trueman Lumber Co., Inc.	St. Leonard, Md.
Schmuck Company, Inc.	Hanover, Pa.	*H. B. Trueman Lumber Company	Huntington, Md.
Schweikart Cedar Supply	Camden, N.J.	Tuckerton Lumber Co.	Surf City, N.J.
Scott Millwork Company, Inc.	Cressona, Pa.	Upsal Lumber & Millwork Co., Inc.	Philadelphia, Pa.
*Seals, Inc.	Elmira, N.Y.	Village Home Center, Inc.	Tannersville, Pa.
The Service Team Home Center	Palmerton, Pa.	Wagner's Mill, Inc.	Rockville, Md.
A. K. Shearer Company	North Wales, Pa.	Walbert Lumber	Mertztown, Pa.
*Shelly & Sons Fenstermacher Co.	Perkasie, Pa.	Waldner Lumber Company	Philadelphia, Pa.
Shelly Enterprises	Souderton, Pa.	Walsh & Company, Inc.	Baltimore, Md.
*Shelly's Building Supply	Kimberton, Pa.	Walter & Jackson, Inc.	Christiana, Pa.
V. J. Sherry	St. Mary's, Pa.	*Walter & Jackson, Inc.	New Holland, Pa.
Shields Lumber & Coal Company	Hockessin, Del.	*Walter & Jackson, Inc.	Quarryville, Pa.
C. A. Shipton, Inc.	Mifflinburg, Pa.	Wardmar Company, Inc.	Edgemont, Pa.
Shone Lumber & Building Materials	Stanton, Del.	E. L. Warren Lumber Company	Port Norris, N.J.
Shrewsbury True Value	Shrewsbury, Pa.	Wattsburg Lumber Company	Wattsburg, Pa.
Simpler Lumber & Coal Company	Felton, Del.	Weaver Lumber & Supply Company	Seneca, Pa.
Simpson Lumber Company	Camden, Del.	Weirton Lumber Company	Weirton, W. Va.
Donald E. Six Building Materials	Middleburg, Md.	Werner Lumber Company	Pine Grove, Pa.
Smick Lumber & Building Center	Quinton, N.J.	West Branch Materials, Inc.	Barnesboro, Pa.
Smith & Reifsnider, Inc.	Westminster, Md.	West Elizabeth Lumber Company	Elizabeth, Pa.
Smith & Richards Lumber Co., Inc.	Bridgeton, N.J.	West Lumber & Building Supply Corp.	Philadelphia, Pa.
Smith Building Supply, Inc.	Churchton, Md.	Wetherill Lumber, Inc.	Bristol, Pa.
*J. C. Snavely & Sons, Inc.	Lancaster, Pa.	Wetherill's, Inc.	Beverly, N.J.
J. C. Snavely & Sons, Inc.	Landisville, Pa.	Wheaton Lumber Company, Inc.	Wheaton, Md.
Estate of George S. Snyder, Inc.	Hatfield, Pa.	Whipple Brothers	Tuckhannock, Pa.
Solomans Home Center, Inc.	Solomans, Md.	Whitesell Brothers Building Supply	Dallas, Pa.
Somers Point Lumber, Inc.	Somers Point, N.J.	Wholesale Express, Inc.	Conshohocken, Pa.
Somerset Door & Column Company	Somerset, Pa.	Wirth Lumberama	Bordentown, N.J.
*SSC Distributors, Inc.	Chester, Pa.	The Wolf Organization, Inc.	York, Pa.
Stanford Home Center	Pittsburgh, Pa.	Phillip Wolf & Son	Lewistown, Pa.
Starke Millwork & Lumber	Easton, Pa.	Wolf Supply Company	Manchester, Pa.
H. M. Stauffer & Sons, Inc.	Leola, Pa.	Woods Brothers Lumber Company	Philadelphia, Pa.
*H. P. Starr & Sons, Inc.	Zelienople, Pa.	York County Lumber Company	York, Pa.
H. P. Starr & Sons, Inc.	Valencia, Pa.	William M. Young Company	Wilmington, Del.
Steckel Concrete Company, Inc.	Phillipsburg, N.J.	Your Building Centers, Inc.	Altoona, Pa.
Stockton Supply Company	Unionville, Pa.	R. A. Zimmerman & Sons, Inc.	Burtonsville, Md.
Stone Harbor Lumber Company	Stone Harbor, N.J.		
Strathmann Lumber Company	Southampton, Pa.	*Branches	
Strathmann-Smith & Boyd	Philadelphia, Pa.		
Stroudsburg Door & Trim, Inc.	Stroudsburg, Pa.		
Lester E. Stuck, Inc.	Mt. Pleasant Mills, Pa.		
Suburban Building Center, Inc.	St. Mary's, Pa.		
*Suburban Lumber Company	Oaklyn, N.J.		
Suburban Window & Door, Inc.	Reading, Pa.		
Sutersville Lumber Company, Inc.	Sutersville, Pa.		
Sutersville Lumber Company, Inc.	Pittsburgh, Pa.		
Sykes-Scholtz-Collins Lumber, Inc.	Philadelphia, Pa.		
Sylvan Build-In	Philadelphia, Pa.		
Szathmary Building Supply Co.	Pleasantville, N.J.		
Tague Lumber, Inc.	Philadelphia, Pa.		
Talbott Lumber Company	Elicott City, Md.		
*Taylor Lumber & Supply Corp.	Rockville, Md.		
Tenth Street Lumber & Millwork, Inc.	Philadelphia, Pa.		
George Ternent Sons	Lonaconing, Md.		

Associate Members

(As of January 1991)
Eastern Building Material Dealers Association

Company Name	City, State
Ace Hardware Corporation	Oak Brook, Ill.
Adam Wholesalers, Inc.	Carlisle, Pa.
Adelman Lumber Company	Pittsburgh, Pa.
Advanced Digital Data, Inc.	Flanders, N.J.
American International	Portland, Oreg.
Andersen Corporation	Bayport, Minn.
Armstrong World Industries Inc.	Hasbrouck Heights, N.J.
Don Aux Associates, Inc.	Malvern, Pa.
Babcock Lumber Company	Pittsburgh Pa.
Bailey, Meyers and Associates	Downington, Pa.
Bangkok Industries, Inc.	Philadelphia, Pa.
Bateman Brothers Lumber	Philadelphia, Pa.
Benco Building Products	Greensburg, Pa.
Bennett Supply Company	Pittsburg, Pa.
Berg Products Corporation	Baltimore, Md.
Bird Roofing Division	Norwood, Mass.
Black Millwork Company, Inc.	Allendale, N.J.
Black River Forest Corporation	Worcester, Pa.
Brady & Sun, Inc.	Worcester, Mass.
Bridgewater Wholesalers, Inc.	Branchburg, N.J.
Paul Brooker Sales	Wichita, Kans.
Browning Metal Products	Peoria, Ill.
Builder Marts of America, Inc.	Greenville, S.C.
Building Material Distributors	Lewisberry, Pa.
Camco Supply, Inc.	Frazer, Pa.
Canadian Forest Products	Conshohocken, Pa.
Caradeo Corporation	Rantoul, Ill.
R. J. Carroll Company, Inc.	Springfield, Pa.
Frank Cerami, Inc.	Jamison, Pa.
P. H. Chadbourne & Company	Bethel, Maine
Clapper's Industries	Meyersdale, Pa.
*E. D. Collier & Son, Inc.	Woodbury, N.J.
Combined Underwriters Inc.	Plymouth Meeting, Pa.
Complete Financial	Rockville, Md.
Component Building Systems Inc.	Newville, Pa.
Compu-Power, Inc.	Columbia, Md.
Computer Applications	Beltsville, Md.
Computer System Dynamics, Inc.	Denver, Colo.
Cook & Dunn Paint Corporation	Carlstadt, N.J.
Corroon & Black/Noyles Services	Media, Pa.
Cotter & Company, Inc.	Fogelsville, Pa.
Crown Lumber Company, Inc.	Baltimore, Md.
DAL National Clearing House	Clifton Heights, Pa.
Dataline Corporation	Wilton, Conn.
Dealers Supply Company	Monessen, Pa.
Delmarva Millwork Corporation	Lancaster, Pa.
Delmarva Sash & Door Company	Barclay, Md.
Derr Lumber & Millwork Company	Bridgeport, Pa.
Derr Flooring Company	Philadelphia, Pa.
Dixie Millwork Company, Inc.	Hagerstown, Md.
Doyle Lumber, Inc.	Martinsville, Va.
Doylestown Ready Mix Concrete	Doylestown, Pa.
Eastern Distributors Company	Philadelphia, Pa.
Eastern Group, Inc.	Newark, Del.
ENAP, Inc.	New Windsor, N.Y.
Energy Associates/Aztec, Inc.	Havertown, Pa.
Engineering Resources	Coopersburg, Pa.
Enterprise Computer Systems	Greenville, S.C.
Erich Brothers	St. Mary's, Pa.
Federal Check Guarantee Corp.	New Castle, Del.
Fessenden Hall	Pennsauken, N.J.
Fessenden Hall of Delaware, Inc.	Dover, Del.
Fessenden Hall of Pennsylvania, Inc.	Lancaster, Pa.
Fessenden Hall, Ltd.	Scranton, Pa.
First State Financial Programs	Wilmington, Del.
Fleet Safety Services Inc.	Navesink, N.J.
Forest-Wise Building District, Inc.	Fruitland, Md.
Furman Lumber, Inc.	Merchantville, N.J.
G & B Composition	Felton, Del.
Georgia Pacific Corporation	Denville, N.J.
Georgia-Pacific	King of Prussia, Pa.
Global Turnkey Systems	Waldwick, N.J.
The Gordon Corporation	Southinton, Conn.
Gro-N-Sell, Inc.	Chalfont, Pa.
Guyon Industries, Inc.	Manheim, Pa.
H/A Building Products, Inc.	Middletown, Pa.
Edward F. Haldeman Associates	Pittsburgh, Pa.
Hallock Lumber Company, Inc.	Maybrooke, N.Y.
R.J. Hanson's Supply, Inc.	Warminster, Pa.
Hardware Wholesalers, Inc.	Fort Wayne, Ind.
Harmonson Stairs	Mt. Laurel, N.J.
Hickson Corporation	Rockville, Md.
Hill & Associates, Inc.	Richmond, Va.
Homasote Company	West Trenton, N.J.
Horstmeier Lumber Company	Baltimore, Md.
K. House & Son Lumber Company	Blossburg, Pa.
Hudson Building Supply	Ashley, Pa.
Husted's, Inc.	Mansfield, Pa.
IKO Manufacturing	Wilmington, Del.
Independence Truck Equipment	Clinton, Md.
Industrial Plywood, Inc.	Reading, Pa.
Information & Financial	Plymouth Meeting, Pa.
Insta-Check	Miami, Fla.
International Wood Products	Queen Anne, Md.
International Wood Productions	Levittown, Pa.
International Wood Products	Milway, N.J.
Iron City Sash & Door Company	Pittsburgh, Pa.
F. Scott Jay & Company, Inc.	Millersville, Md.
KSI Building Products, Inc.	Cobleskill, N.Y.
Kern Distributing Company	Jessup, Md.
Keyline Company, Inc.	Exton, Pa.
Keystone Wood Treating Corp.	Oxford, Pa.
Keystone Millwork	Zionville, Pa.
Kohl Building Products	Reading, Pa.
Kuhns Bros. Log Homes, Inc.	Lewisburg, Pa.
Kuntz Lesher Siegrist	Lancaster, Pa.
*Russell Kuntz	Freedom, Pa.
Lambton Manufacturing Limited	Maple Glen, Pa.
Larson Manufacturing Company	Allentown, Pa.
Level Line, Inc.	Lakewood, N.J.
Lumber Insurance Company	Framingham, Ma.
Lumbermen Associates, Inc.	Bristol, Pa.
Lumbermens Merchandising Corp.	Wayne, Pa.
M & R Products	Vineland, N.J.

Magee-Fine Lumber Company	Pottstown, Pa.
Mala, Inc.	Media, Pa.
Watson Malone & Sons, Inc.	Haverford, Pa.
Manufacturers Reserve	Irvington, N.J.
Marketech Marketing Services	Lancaster, Pa.
MCS, Inc.	Pittsburgh, Pa.
R. S. Means Company, Inc.	Kingston, Mass.
Merrill Lumber Company, Inc.	Philadelphia, Pa.
Metropolitan Industries, Inc.	Hyattsville, Md.
Mid-State Lumber Company	Branchburg, N.J.
Mid-State Lumber Company	Kingston, Pa.
Midland Hardware Distributors	Oakland, N.J.
C. H. Miller Hardware	Huntingdon, Pa.
Moore Business Forms	Niagara Falls, N.Y.
Edward J. Moran Lumber Lumber	West Chester, Pa.
Morgan Distribution Company	Mechanicsburg, Pa.
Morgan Technologies	Columbia, Md.
The Moulding & Millwork	Baltimore, Md.
Robert R. Mueller, Ltd.	Wayne, Pa.
Musser's House of Carpets	Goodville, Pa.
MW Manufacturing, Inc.	Rocky Mount, Va.
National Nail Corporation	Syracuse, N.Y.
National Wood Preservers, Inc.	Havertown, Pa.
Needham & Associates, Inc.	Pittsburgh, Pa.
F. J. Newmeyer Lumber Company	Rahway, N.J.
North East Hardwoods, Inc.	Mt. Jewett, Pa.
*Northeast Drywall Supply Co.	Wilkes-Barre, Pa.
Northeast Sales Corp.	Piscataway, N.J.
Off The Wall Company, Inc.	Telford, Pa.
Osmose-Wood Preserving	Griffin, Ga.
Owens-Corning Fiberglass	Wayne, Pa.
Pacific Mutual Door Company	Baltimore, Md.
Pan American Building Material	Cornwell Heights, Pa.
PBR Financial Services, Inc.	Elkins Park, Pa.
Peachtree Doors, Inc.	Norcross, Ga.
Pearce Fireproof Company, Inc.	Huntington Valley, Pa.
Penn-Atlantic Millwork, Inc.	Philadelphia, Pa.
Pennsylvania Lumbermens Mutual	Philadelphia, Pa.
Philadelphia Reserve Supply	Croydon, Pa.
Plunkett-Webster, Inc.	South Plainfield, N.J.
Ply Gems Industry	Philadelphia, Pa.
The PMA Group	Lemoyne, Pa.
Potomac Supply Corporation	Kinsale, Va.
Pre-Mix Industries, Inc.	Annadale, Va.
Preferred Marketing Associates	Wenonah, N.J.
Prudential LMI Commercial	Mansfield, Ohio
*Quality Wood Treating Company	Mars, Pa.
R & R Wood Products, Inc.	Mainland, Pa.
Red Rose Building Systems, Inc.	Stevens, Pa.
Redi-Truss, Inc.	Salisbury, Md.
Reeb Millwork Corporation	Bethlehem, Pa.
Reid & Wright, Inc.	Arcata, Calif.
Remodel America, Inc.	Willow Grove, Pa.
Roland & Roland, Inc.	Bethlehem, Pa.
Roseville Woodworks, Inc.	Mansfield, Pa.
Alvin Rothenberger, Jr., Inc.	Worcester, Pa.
Russell Plywood, Inc.	Reading, Pa.
*Russin Lumber Corporation	Montgomery, N.Y.
Saul, Ewing, Remick & Saul	Philadelphia, Pa.
Schmoyers Home Center	Boyertown, Pa.
Scholl Lumber Company	Bethlehem, Pa.
Seaboard International Lumber	Jericho, N.Y.
Senco Products, Inc.	Richmond, Va.
Servistar Corporation	Butler, Pa.
John N. Serwo & Associates	Southampton, Pa.
Sherwood Lumber Corporation	Islandia, N.Y.
Shuster's Building Components	Irwin, Pa.
Simpson Strong-Tie Company, Inc.	Orwigsburg, Pa.
Slabaugh Custom Stairs, Ltd.	Quakertown, Pa.
Snavely Forest Products, Inc.	Pittsburgh, Pa.
*Snavely Forest Products Corp.	Baltimore, Md.
*Snavely Forest Products Corp.	Freehold, N.J.
Soult Wholesale Company	Clearfield, Pa.
South Atlantic Wood	Elizabeth City, N.C.
Specialty Building Products	Creamery, Pa.
Spruce Computer Systems	Latham, N.Y.
St. Mary's Precision Homes, Inc.	St. Mary's, Pa.
Stanton Door Company	Exton, Pa.
Stanton Supply	Newark, Del.
State Mutual Life Assurance co.	Worcester, Mass.
Stockton Bates & Company	Hershey, Pa.
Stop-Loss Associates, Inc.	Framingham, Mass.
Storage Concepts	Ellicott City, Md.
Swartz Supply Company, Inc.	Harrisburg, Pa.
Sylvania Wood Products Company	Philadelphia, Pa.
*Sylvania Wood Products Company	Langhorne, Pa.
Systems & Software Services	Media, Pa.
Tei Construction Fabrics	Baltimore, Md.
Thermo-Vu Sunlite Industries	Edgewood, N.Y.
William F. Thesing Mfg. Co., Inc.	Willow Grove, Pa.
Timbermark, Inc.	Hershey, Pa.
Toll Intergrated Systems	Morrisville, Pa.
Town & Country Kitchen	Reinholds, Pa.
Tri-State Millwork Company	Bellmawr, N.J.
Triad Systems Corporation	Livermore, Calif.
Trimline Windows, Inc.	Ivyland, Pa.
Trinity Lams of New Jersey, Inc.	Brick, N.J.
Trumbell Industries	Warren, Ohio
Trusco, Inc.	Murrysville, Pa.
Universal Business Systems, Inc.	Somerville, N.J.
Universal Forest Procuts, Inc.	Ranson, W. Va.
Universal Suppliers, Inc.	Sellinsgrove, Pa.
US Truck Craines, Inc.	York, Pa.
Velux-America	Langhorne, Pa.
Versyss, Inc.	Valley Forge, Pa.
A. C. Waddell & Associates, Ltd.	Jamestown, N.Y.
Wasco Products, Inc.	Dallas, Tex.
Walter H. Weaber Sons, Inc.	Lebanon, Pa.
Weather Sheild Mgf., Inc.	Camp Hill, Pa.
Wholesale Lumber Services, Inc.	Mt. Joy, Pa.
Martin Wiegand, Inc.	Washington, D.C.
Wiener & Crowley, Inc.	Leonia, N.J.
George L. Wilson Company, Inc.	Pittsburg, Pa.
Win-Kit Company	Reinholds, Pa.
V. P. Winter Distributing	Paxinos, Pa.
*V. P. Winter Company	Bridgeville, Del.
*V. P. Winter Company	Williamsport, Md.
Wood Treating Corp. of Burlington	Burlington, N.J.
The James Wood Company	Williamsport, Pa.
Woodhouse Post & Beam	Mansfield, Pa.
Woxall Woodcraft	Green Lane, Pa.

*Branches

1991

Nominating Committee *ex officio*

Mr. Aldo Braido, CHRM
General Supply Company
620 Lehigh Drive
P.O. Box 191
Easton, PA 18042

Mr. Perry E. Brunk
People's Supply Co., Inc.
3200 Kennilworth Avenue
Hyattsville, MD 20781

Mr. Robert M. Bushey
Cavetown Planing Mill Co.
Rt. 66, Box 51
Cavetown, MD 21720

Mr. Jay F. Risser
J. H. Brubaker, Inc.
2008 Marietta Avenue
Lancaster, PA 17603

Mr. J. Fred Robinson
Newark Lumber Company
221 E. Main Street
Newark, DE 19715

Mr. Karl J. Westover
Allensville Planning Mill
108 E. Main Street
Allensville PA 17002

Mr. Edwin F. Scholtz
Sykes-Scholtz-Collins, Lumber
Clarissa & Bristol Streets
Phildelphia, PA 19140

Mr. William E. Shone, Jr.
Shone Lumber & Building Materials
P.O. Box 6007
Stanton, DE 19804

Mr. Vincent J. Tague
Tague Lumber, Inc.
Belfield Ave & High St.
Philadelphia, PA 19144

Mr. David Waitz
Emily Lumber Company
2115-25 S. 8th Street
Philadelphia, PA 19148

Mr. James P. Rauch*
Crafton Lumber & Supply
Nobel & Bradford Avenues
P.O. Box 4443
Pittsburgh, PA 15205-0443

Mr. Bruce Ferretti *
Lehigh Lumber Company
6410 Airport Road
Bethlehem, PA 18017

Mr. David B. Kreidler *

1991–92

Bylaws Committee *ex officio*

Mr. B. Harold Smick, Jr.
Chariman
Smick Lumber & Building Materials
Route 49, P.O. Box 127
Quinton, NJ 08072

Mr. Charles M. Bowers
Wm. D. Bowers Lumber Co.
10620 Woodsboro Pike
Woodsboro, MD 21798

Mr. Howard B. Dries
Dries-Do-It-Center
3500 Brookside Road
Macungie, PA 19062

Mr. John H. Eaton, Jr.
Barrons Enterprises
23 W. Diamond Avenue
Gaithersburg, MD 20877

Mr. James P. Rauch *
Crafton Lumber & Supply
Nobel & Bradford Avenues
P.O. Box 4443
Pittsburgh, PA 15205-0443

Mr. Bruce Ferretti*
Lehigh Lumber Company
6410 Airport Road
Bethlehem, PA 18017

Mr. Gerald S. Greene
Christy's Supplies
Route 70 Box 157
Medford, NJ 08055-0157

Mr. William D. Hayes
Moxham Lumber Company
Park Ave. & Dupont St.
Johnstown, PA 15902

Mr. Dale W. Parker
Cramer's Cashway, Inc.
320 N. Courtland St.
E. Stroudsburg, PA 18301

Mr. Gregory Shelly
Shelly & Fenstermacher
64 Highland Avenue
Souderton, PA 18963

Mr. Karl J. Westover
Allensville Planning Mill
108 E. Main Street
Allensville, PA 17002

Mr. David B. Kreidler*

1991-92

Legislative Committee *ex officio*

Mr. Terry L. Kauffman
Co-Chairman
Reinholds Lumber Millwork, Inc.
150 Lincoln Ave. Box 209
Reinholds, PA 17569

Mr. John F. Carroll, Jr.
Heintz Home Center
255 Wilmington Pike
Chadds Ford, PA 19317

Mr. Sylvan Cohen
Sylvan Build-In
5707 Walnut Street
Philadelphia, PA 19139

Mr. Edward H. Davis, Jr.
Moulton H. Davis Estate
1564 McDaniel Drive
Westtown Business Center
West Chester, PA 19380

Mr. John M. Denlinger
Denlinger, Inc.
Box 369
Paradise, PA 17562

Mr. O. Grant Little
Little Lumber Company
Box L
Benton, PA 17814

Mr. R. Brian Shober
NELCO
2nd & State Streets
Hamburg, PA 19526

Mr. H. Paul Starr
H. P. Starr & Sons, Inc.
R.D. 3, Old Plank Road
Valencia, PA 16059

Mr. Vincent J. Tague
Tague Lumber, Inc.
Belfield Ave. & High St.
Philadelphia, PA 19144

Mr. James P. Rauch *
Crafton Lumber & Supply
Noble & Bradford Avenues
P.O. Box 4443
Pittsburgh, PA 15205-0443

Mr. David Waitz
Emily Lumber Company
2115-25 So. 8th Street
Philadelphia, PA 19148

Mr. B. Harold Smick, Jr.
Co-Chairman
Smick Lumber & Building Material
Route 49, P.O. Box 127
Quinton, NJ 08072

Mr. Harry Blair, Jr.
C. A. Niece Lumber Co.
N. Union & Elm Streets
Lambertville, NJ 08242

Mr. Asa L. Colson III
Colson's Home & Building Center
18th & New Jersey Ave.
North Wildwood, NJ 08242

Mr. Gene S. DiMedio
Dubell Lumber Company
102 Route 73
P.O. Box 160
Cedar Brook, NJ 08010

Mr. Gerald S. Greene
Christy's Supplies
Route 70, Box 157
Medford, NJ 08055-0157

Mr. J. Fred Robinson
Co-Chairman
Newark Lumber Company
221 E. Main St., Box 417
Newark, DE 19711

Mr. Farrell L. Goble
Brosius-Eliason Co.
508 South Street
New Castle, DE 19720

Mr. Lee R. Harman
U. L. Harman, Inc.
Main Street, Box 56
Marydel, DE 21649

Mr. John E. Smith, Jr.
Smith Building Supply, Inc.
5573 Shady Side Road
Churchton, MD 20733

Mr. Robert M. Bushey
Co-Chairman
Cavetown Planing Mill
Box 51
Cavetown, MD 21720

Mr. Charles M. Bowers
Wm. D. Bowers Lumber Co.
10620 Woodsboro Pike
Woodsboro, MD 20781

Mr. Perry Brunk
Peoples Supply Company
3200 Kenilworth Avenue
Hyattsville, MD 20781

Ms. Peggy Bushey
Cavetown Planing Mill
Box 52
Cavetown, MD 21720

Mrs. Annabelle Hornsby
Allied Building Center, Inc.
Moss Hill Lane
Salisbury, MD 21801

Mr. James D. Kolker
Maryland Lumber Company
2601 W. Franklin Street
Baltimore, MD 21223

Mr. W. Russell Lamar
Lamar & Wallace, Inc.
7000 Old Landover Road
Landover, MD 20785-1595

Mr. Van T. Mitchell
MSI, Inc.
6 St. Mary's Place
P.O. Box W
LaPlata, MD 20646

Mr. James D. Neal
Nuttle Lumber Company
P. O. Box 475
Denton, MD 20646

Mr. Bruce Ferretti *
Lehigh Lumber Company
6410 Airport Road
Bethlehem, PA 18017

Mr. David B. Kreidler *

1991–92

Education Committee *ex officio

Mr. John H. Eaton, Jr.,
Chairman
Barrons Enterprises, Inc.
23 W. Diamond Street
Gaithersburg, MD 20877

Mr. Robert Albrecht
Timbermark, Inc.
P.O. Box 114
Hershey, PA 17033

Mr. John F. Beaver
Roland & Roland, Inc.
698 Angenese Street
Harrisburg, PA 17710

Mr. Arthur R. Borden
Lewisburg Bldrs Supply
Rt 15 & Hafer Road
P.O. Box 349
Lewisburg, PA 17837

Mr. Jon C. Clapper
Clapper's Bldg Mat'ls, Inc.
Eleventh Avenue
P.O. Box 335
Meyersdale, PA 15552

Mr. Michael Carty
Butz Building Centers
Sand Quarry Road
Palmerton, PA 18071

Mr. Charles Davis
Barrons Enterprises, Inc.
23 W. Diamond Street
Gaithersburg, MD 20877

Mr. John M. Denlinger
Denlinger, Inc.
Box 369
Paradise, PA 17562

Mr. Gene S. DiMedio
Dubell Lumber Company
102 Route 73
P.O. Box 160
Cedar Brook, NJ 08010

Mr. Daniel Duerring
Stanford Home Center
2001 Route 286
Pittsburgh, PA 15239

Mr. I. S. Eberly
Eberly Lumber Company
High Street
Mechanicsburg, PA 17055

Mr. Paul Fatur
Babcock Lumber Co.
2220 Palmer Street
Pittsburgh, PA 15218

Mr. Lawrence Forman
H. B. Trueman Lumber Co.
P.O. Box 118
St. Leonard, MD 20685

Mr. Roland L. Green, Jr.
Hyatt Building Supply Co., Inc.
26200 Ridge Road
Damascus, MD 20872

Mr. W. James Hollenbach
W. Hollenbach Company
P.O. Box 507
Boyertown, PA 19512

Ms. Annabelle M. Hornsby
Allied Building Center, Inc.
Moss Hill Lane
P.O. Box 2535
Salisbury, MD 21801

Mr. James D. Kolker
Maryland Lumber Company
2601 W. Franklin Street
Baltimore, MD 21223

Mr. William J. Lane
Brosius-Eliason Co.
508 South Street
New Castle, DE 19720

Mr. Brian J. Lucas
Gilbert Lumber & Supply Co.
3711 Walnut Street
McKeesport, PA 15132

Mr. James J. Maloney, Jr.
Sykes-Scholtz-Collins Library
Clarissa & Bristol Streets
Philadelphia, PA 19140

Mr. Joseph D. Marsilio
Hazle Builders, Inc.
30th & North Church St.
Hazleton, PA 18201

Mr. David Maust
Casselman Lumber
Star Route Box 32B
Grantsville, MD 21536

Mr. Steven D. Mitchell
G. R. Mitchell, Inc.
Refton Road, Box 128
Refton, PA 17568

Mr. Van T. Mitchell
MSI, Inc.
6 St. Mary's Place
P.O. Box W
LaPlata, MD 20646

Mr. Domenic J. Pileggi, Jr.
Concord Plywood & Supply
Upland & Kurlin Streets
Upland, PA 19013

Mr. Philip Skarada
Your Building Center
2607 Beale Avenue
Box 1230
Altoona, PA 16603-1230

Mr. Andrew Williams
H. O. P. Lumber Co.
1331 Freeport Road
Pittsburgh, PA 15238

Mr. James P. Rauch*
Crafton Lumber & Supply
Noble & Bradford Avenues
P.O. Box 4443
Pittsburgh PA 15205-0443

Mr. Bruce Ferretti*
Lehigh Lumber Company
6410 Airport Road
Bethlehem, PA 18017

Mr. David B. Kreidler*

1991-92

Business Development Committee *ex officio*

Mr. Gregory Shelly, Chairman
Shelly & Fenstermacher
64 Highland Avenue
Souderton, PA 18969

Mr. Harry Blair, Jr.
C. A. Niece Co., Inc.
N. Union & Elm Streets
P.O. Box 68
Lambertville, NJ 08530

Mr. Sylvan Cohen
Sylvan Build-In Sykes
5707 Walnut Street
Philadelphia, PA 19139

Mr. Leonard Erbe
C.H. Marshall, Inc.
Media Station Road
Box 196
Media, PA 19063

Mr. Farrell L. Goble
Brosius Eliason Company
508 South Street
New Castle, DE 19720

Mr. Gerald S. Greene
Christy's Supplies
Route 70, Box 157
Medford, NJ 08055-0157

Mr. W. James Hollenbach
W. Hollenbach Company
P.O. Box 507
Boyertown, PA 19512

Mr. Matthew H. Kohn
Allentown Bethlehem
Lumber Company
125 Union Blvd
Allentown, PA 18103

Mr. W. Russell Lamar, Jr.
Lamar & Wallace, Inc.
7000 Old Landover Rd.
Landover, MD 20785-1595

Mr. Bruce Nelson
Tuckerton Lumber
200 North Blvd
P.O. Box 370
Surf City, NJ 08087

Mr. Brian J. Lucas
Gilbert Lumber Company
3711 Walnut Street
McKeesport, PA 15132

Mr. G. Eugene Mackie
Mackies Home Center
171 W. Main Street
P.O. Box 549
Cecilton, MD 21913

Mr. James J. Maloney, Jr.
Scholtz Collins Lumber
Clarissa & Bristol Streets
Philadelphia, PA 19140

Mr. Howard Metzler
Cobilt, Inc.
115 W. Penn Street
Martinsburg, PA 16662

Mr. J. Kirk Miller
Sutersville Lumber co., Inc.
Eighth Ave., P.O. Box 408
Sutersville, PA 15083

Mr. Fred Musselman
Musselman Lumber, Inc.
300 Brimmer Avenue
New Holland, PA 17557

Mr. James D. Neal
The Nuttle Lumber Co.
P.O. Box 475
Denton, MD 21629

Mr. G. Robert Overhiser
Collingdale Millwork Co.
Sharon & Pusey Avenue
Box 385
Collingdale, PA 19023

Mr. David G. Patterson
Patterson Lumber Company
41-45 Charleston Street
Wellsboro, PA 16901

Mr. Willis R. Meyer
Suburban Building Center, Inc.
Johnsonburg Road
St. Mary's, PA 15857

Mr. Jay F. Risser
J. H. Brubaker, Inc.
2008 Marietta Avenue
Lancaster, PA 17603

Mr. R. Brian Shober
NELCO
2nd & State Streets
Hamburg, PA 19526

Mr. Philip Skarada
Your Building Center
2607 Beale Avenue
Box 1230
Altoona, PA 16603-1230

Mr. William K. Turner
E. S. Adkins & Company
801 N. Salisbury Blvd.
P.O. Box 1779
Salisbury, MD 21801

Mr. Robert B. Walter, III
Walter & Jackson, Inc.
Box 160
Christiana, PA 17509

Mr. Andrew Williams
H. O. P. Lumber Co.
1331 Freeport Road
Pittsburgh, PA 15238

Mr. James P. Rauch*
Crafton Lumber & Supply
Noble & Bradford Avenues
P.O. Box 4443
Pittsburgh, PA 15205-0443

Mr. Bruce Ferretti*
Lehigh Lumber Company
6410 Airport Road
Bethlehem, PA 18017

Mr. David B. Kreidler*

Mr. Daniel Duerring
Stanford Home Center
2001 Route 286
Pittsburgh, PA 15239

Mr. Frank Kane
Kane Lumber, Inc.
500 W. College Avenue
Tamaqua, PA 18252

1991-92

Membership Development Committee *ex officio*

Mr. O. Grant Little
Chairman
Little Lumber Company
Box L
Benton, PA 17814

Mr. John F. Carroll, Jr.
Heintz Home Center
255 Wilmington Pike
Chadds Ford, PA 19144

Mr. Howard B. Dries
Dries-Do-It-Center
3580 Brookside Road
Macungie, PA 18062

Mr. Roland L. Green, Jr.
Hyatt Bldg Supply Co., Inc.
26200 Ridge Road
Damascus, MD 20872

Mr. Lee R. Harman
U. L. Harman, Inc.
1 Main Street, Box 56
Marydel, DE 19964

Mr. William D. Hayes
Moxham Lumber Company
Park Avenue & Dupont St.
Johnstown, PA 15902

Mr. John R. Shone
Shone Lumber & Building Material
P.O. Box 6007
Stanton, DE 19804

Mr. Paul Starr
H. P. Starr & Sons, Inc.
Glade Mills, R.D. 2
Valencia, PA 16059

Mr. William K. Turner
E. S. Adkins & Company
801 N. Salisbury Blvd
P.O. Box 1779
Salisbury, MD 21801

Mr. Edward Kostick
Middletown Lumber Inc.
Brown & Clinton Streets
Middletown, PA 17057

Mr. Daniel Kauffman
Otto & Hollinger, Inc.
418 Market Street
P.O. Box 14
Lemoyne, PA 19043

Mr. James P. Rauch*
Crafton Lumber & Supply
Noble & Bradford Avenues
P.O. Box 4443
Pittsburgh, PA 15205-0443

Mr. Bruce Ferretti*
Lehigh Lumber Company
6410 Airport Road
Bethlehem, PA 18017

Mr. Steven D. Mitchell
G. R. Mitchell, Inc.
Refton Road, Box 128
Refton, PA 17568

Mr. G. Robert Overhiser
Collingale Millwork Co.
P.O. Box 385
Collingdale, PA 19023

Mr. John K. Runnette, Jr.
Manufacturers Reserve Supply Company
16 Woolsey Street
Irvington, NJ 07111

Mr. David B. Kreidler*

1991-92

Membership Development Program Committee *ex officio*

Mr. Randall Brunk, Chairman
Peoples Supply Co., Inc.
3200 Kennilworth Ave.
Hyattsville, MD 20781

Mr. Gary Allshouse
Moulton H. Davis Lbr. Co.
515 S. Bolmar Street
West Chester, PA 19382

Mr. Douglas S. Bomberger
Long & Bomberger
45 N. Broad Street
Lititz, PA 17543

Mr. John F. Carroll, Jr.
Heintz Home Center
255 Wilmington Pike
Chadds Ford, PA 19317

Mr. Chip Chapin
Chapin Lumber & Supply Co
5th St. Highway and Tuckerton Road
Temple, PA 19560

Mr. Christopher L. Smith
Watson Malone & Sons, Inc.
Box 176
Haverford, PA 19041

Mr. J. Douglas Wetherill
Wetherill's, Inc.
Route 130
Beverly, NJ 08010

Mr. John H. Rearick III
Vice Chairman
J. H. Rearick & Son, Inc.
98 West Church Street
Dillsburg, PA 17019

Ms. Elizabeth Griffin
Lewisburg Builders Supply
Route 15 & Hafer Road
P.O. Box 349
Lewisburg, PA 17837

Mr. James P. Rauch *
Crafton Lumber & Supply
Noble & Bradford Avenues
P.O. Box 4443
Pittsburgh, PA 15205-0443

Mr. Bruce Ferretti *
Lehigh Lumber Company
6410 Airport Road
Bethlehem, PA 18017

Mr. Daniel Heckler
Est. of Geo. S. Snyder
1700 Hatfield Valley Rd
Hatfield, PA 19440

Ms. Donna Hankins
H. H. Hankins & Bro.
12 W. Broad Street
Bridgeton, NJ 08302

Mr. David B. Kreidler*

1991-92

Centennial Committee *ex officio

Mr. B. Harold Smick
Chairman
Smick Lumber & Building Materials
Route 49, P.O. Box 127
Quinton, NJ 08072

Mr. Farrell L. Goble
Brosius-Eliason Company
508 South Street
New Castle, DE 19720

Mr. Frank M. Hankins, Jr.
H. H. Hankins & Bro.
12 W. Broad Street
P.O. Box 498
Bridgeton, NJ 08302-0388

Mr. Lee R. Harman
U. L. Harman, Inc.
1 Main Street
P.O. Box 56
Marydel, DE 19964

Mrs. Annabelle Hornsby
Allied Building Center
Moss Hill Lane
P.O. Box 2535
Salisbury, MD 21801

Mr. James J. Maloney, Jr.
Sykes Scholtz Collins, Lumber
Clarissa & Bristol Sts.
Philadelphia, PA 19140

Ms. Alta M. Miller
Peachey Builders
R.D. No. 1, Box 272 B
Belleville, PA 17004

Mrs. Mary K. Rearick
J. H. Rearick & Son, Inc.
98 W. Church Street
Dillsburg, PA 17019

Mr. Thomas F. Rider
Faxon Lumber Company
1700 East 3rd Street
P.O. Box 1027
Williamsport, PA 17703-1027

Mr. Arthur R. Borden
Lewisburg Bldrs Supply
Rt. 15 & Hafer Road
Lewisburg, PA 17837

Mr. A. K. Shearer III
A. K. Shearer Company
123 S. Second Street
P.O. Box 1039
North Wales, PA 19454

Mr. William E. Shone, Jr.
Shone Lumber & Building
P.O. Box 6007
Stanton, DE 19804

Mr. John E. Smith, Jr.
Smith Building Supply
5573 Shady Side Road
Churchton, MD 20733

Mr. William A. Smith, Jr.
Middlesex Building & Roofing Supply
State St. P.O. Box 223
Middlesex, DE 19966

Mr. Vincent J. Tague
Tague Lumber, Inc.
Belfield Ave. & High Sts.
Philadelphia, PA 19144

Mr. Claude S. Wetherill, 3rd
Wetherill Lumber, Inc.
3230 Bath Road
Bristol, PA 19007

Mr. James P. Rauch *
Crafton Lumber & Supply
Noble & Bradford Ave.
P.O. Box 443
Pittsburgh, PA 15205

Mr. Bruce C. Ferretti *
Lehigh Lumber Company
6410 Airport Road
Bethlehem, PA 18017

Mr. Joseph D. Marsilio
Hazle Builders, Inc.
30th & N. Church Street
Hazleton, PA 18201

Mr. J. Fred Robinson
Newark Lumber Company
221 East Main Street
P.O. Box 417
Newark, DE 19715

Mr. John H. Auld
John H. Auld & Bro. Co.
3919 Rt 8 & Duncan
Allison Park, PA 15101

Mr. David Birchmire
Deepwater Lumber & Supply
376 N. Broadway
Pennsville, NJ 08070

Mr. Robert M. Bushey
Cavetown Planing Mill
Route 66, P.O. Box 51
Cavetown, MD 21720

Mr. Frank Braceland
701 Larchwood Lane
Villanova, PA 19085

Mr. Charles C. Cluss
O. C. Cluss Lumber Co.
South Pennsylvania Avenue
Box 696
Uniontown, PA 15406

Mr. Leonard Desmet
Desmet Lumber & Supply Co.
Box 306
Cecil, PA 15321

Mr. John W. Eckman
Eckman Lumber Co., Inc.
1280 Main Road
Lehighton, PA 18235

David B. Kreidler *
Harry H. Johnson III

1991

Convention Advisory Committee *ex officio*

Mr. Vincent J. Tague,
Chairman
Tague Lumber Company
Belfield Ave. & High St.
Philadelphia PA 19144

Mr. Terry L. Kauffman
Reinholds Lbr & Millwork
150 Lincoln Avenue
P.O. Box 209
Reinholds, PA 17569

Mr. W. Russell Lamar, Jr.
Lamar & Wallace, Inc.
7000 Old Landover Road
Landover, MD 20785-1595

Mr. G. Eugene Mackie
Mackie's Home Center
171 W. Main Street
P.O. Box 549
Cecilton, MD 21913

Mr. J. Fred Robinson
Newark Lumber Company
221 East Main Street
P.O. Box 417
Newark, DE 19715

Mr. John H. Eaton, Jr.
Barrons Enterprises
23 W. Diamond Avenue
Gaithersburg, MD 20877

Mr. H. Paul Starr
H. P. Starr & Sons, Inc.
Glade Mills, R.D. 2
Valencia, PA 16059

Mr. James D. Kolker
Maryland Lumber Company
2601 W. Franklin Street
Baltimore, MD 21223

Mr. David G. Patterson
Patterson Lumber Company
41 N. 45 Charleston Street
Wellsboro, PA 16901

Mr. William Wallace
Preferred Marketing Assoc.
Box 148
Wenonah, NJ 08090

Mr. Edwin F. Scholtz
Sykes-Scholtz-Collins Lbr.
Clarissa & Bristol Sts.
Philadelphia, PA 19140

Mr. Dale C. Adams
Benco Building Products
53 Eby Chiques Road
Mt. Joy, PA 17601

Mr. William E. Baer, Jr.
Eastern Distributors Co.
34th St. & Indiana Ave.
Philadelphia, PA 19132

Mr. John F. Beaver
Roland & Roland, Inc.
698 Angenese Street
Harrisburg, PA 17110

Mr. Bernard Bernstein
Mid-State Lumber Corp.
200 Industrial Parkway
Branchburg, NJ 08876

Mr. Ray Bures, Jr.
H. M. Stauffer & Sons, Inc.
33 Glenola Drive
P.O. Box 38
Leola, PA 17540

Mr. Thomas Fitzgerald
Ply-Gem Distribution
E Street & Erie Ave.
P.O. Box 26928
Philadelphia, PA 19134

Mr. Daniel Flynn
Delmarva Millwork Corp.
1275 S. Manheim Pike
P.O. Box 4068
Lancaster, PA 17604

Ms. Denise J. Gilmartin
CAS, Inc.
4715 Sellman Road
Beltsville, MD 20705

Mr. Geoffrey A. Jones
NORCO Windows of Pennsylvania
175 Lions Drive
Valmont Industrial Park
West Hazleton, PA 18210

Mr. Frank Kelly
Adams Wholesalers
Chestnut St. & Academy Ave.
Woodbury Heights, NJ 08096

Mr. James C. Krebs
Morgan Distribution
Newark Distribution Center
101 Interchange Blvd.
Newark, DE 19711

Mr. Barry Mursky
Thermo-Vu Sunlight Ind.
51 Rodeo Drive
Edgewood, NY 11717

Mr. Steven Boyd
Manufacturers Reserve
Supply, Inc.
16 Woolsey Street
Irvington, NJ 07111

Mr. James Rodgers
Cook & Dunn Paint Corp
700 Gotham Parkway
P.O. Box 836
Carlstadt, NJ 07072

Mr. Timothy H. Sylvester
Delmarva Sash & Door Co.
of Maryland, Inc.
P.O. Box 128
Barclay, MD 21607

Mr. Raymond Weitzel
Win-Kit Company
P.O. Box 90
Reinholds, PA 17569

Mr. James P. Rauch *
Mr. Bruce C. Ferretti *
Mr. David B. Kreidler *

1991

Associate Member Committee *ex officio*

Mr. James P. Rauch, Chrm
Crafton Lumber Company
Noble & Bradford Avenues
P.O. Box 4443
Pittsburgh, PA 15205-0443

Mr. Dale Adams
Benco Bldg Products, Inc.
1335 South Main Street
P.O. Box 728
Greensburg, PA 15601

Mr. William E. Baer, Jr.
Eastern Distributor Co.
34th & Indiana Avenue
Philadelphia, PA 19132

Mr. Bernard Bernstein
Mid-State Lumber Corp.
200 Industrial Parkway
Branchburg, NJ 08876

Mr. James Bounds
Delmarva Millwork Corp.
1127 Manheim Pike
P.O. Box 4068
Lancaster, PA 17604

Mr. Herbert W. Carden
Potomac Supply Corporation
Route 203 North
Kinsale, VA 22488

Mr. Edgar Lobley
Fessenden Hall of PA, Inc.
3021 Industry Drive
Lancaster, PA 17603

Mr. Thomas Deegan
Lumbermen Associates Inc.
840 Cottman Avenue
P.O. Box 15375
Philadelphia, PA 19111

Mr. Russell DiGiallorenzo
Plywood, Inc. V. P.
401 Old Wyomissing Road
Reading, PA 19611

Mr. Paul Fatur
Babcock Lumber Company
2220 Palmer Street
Pittsburgh, PA 15218

Mr. Robert A. Jones
Adams Wholesalers
Chestnut St & Academy
Woodbury Heights, NJ 08096

Mr. John R. Kohl
Kohl Building Products
P.O. Box 14746
Reading, PA 19612

Mr. Robert C. O'Neil
Morgan Distributors
P.O. Box 2003
Mechanicsburg, PA 17055

Mr. Joe Palencar, Sr.
The Moulding & Millwork
Company, Inc.
5805 Southwestern Blvd
Baltimore, MD 21227

Mr. Mike Parli
Manufacturers Reserve Sup
1600 Woolsey Street
Irvington, NJ 07111

Mr. Frank Schaefer
Pan American Building Materials
2901 Samuel Drive
Cornwells Heights, PA 19020

Mr. Carl D. Shaffer
Quality Wood Treating
217 Executive Drive
Suite 202
Mars, PA 16046

Mr. Ray Thomas Russell
Winter Distributing Co
P.O. Box 180
Paxinos, PA 17860

Mr. William P. Wallace
Preferred Marketing Assoc.
P.O. Box 148
Wenonah, NJ 08090

Mr. Charles Weitzel
Win Kit Company
P.O. Box 90, Route 897
Reinholds, PA 17569

Mr. Martin Wiegand
Martin Wiegand, Inc.
6000 Chillum Place, N.E.
Washington, D.C. 20011

Mr. Norman Wolff
Adelman Lumber Company
13th & Smallman Streets
Pittsburgh, PA 15222

Mr. James P. Rauch *
Crafton Lumber & Supply
Noble & Bradford Avenue
P.O. Box 4443
Pittsburgh, PA 15205-0443

Mr. Bruce C. Ferretti *
Lehigh Lumber Company
6410 Airport Road
Bethlehem, PA 18017

Mr. David B. Kreidler *

EBMDA Senior Staff. From left to right seated: Richard W. Brown, Gary W. Zook, Anna A. Grandizio. From left to right standing: E. Thomas Fleck, Jonas Green, Jr., Joseph C. Bradley, Jr., and Robert E. Clark.
Photo courtesy of EBMDA

EBMDA Staff. From left to right seated: Mary Ann Brady, Mary Bennett, Eileen Stambaugh, Mary Hamilton, and Nancy Hooper. From left to right standing: Michelle Newcomb, Ada O'Connor, Clara Boughner, Carol Benetez, Linda Linn, Dennis Chupein, Marietta Kavanaugh, Nancy Finley, and Nicole Ferrara. Photo courtesy of EBMDA

1991 EBMDA Staff

100 YEARS *Celebration* BUYING SHOW

FEBRUARY 4, 5, 6, 1992
LARGEST BUYING SHOW
OF THIS TYPE IN THE
MID-ATLANTIC STATES!

STARTING A NEW CENTURY OF SERVICE AND SUCCESS!

Taj Mahal
Atlantic City
New Jersey

REMEMBER THE PAST   HONOR THE PRESENT   LOOK FORWARD TO THE FUTURE

TAJ MAHAL, ATLANTIC CITY, NEW JERSEY

EASTERN BUILDING MATERIAL DEALERS ASSOCIATION

This convention and exhibition is the 100th Annual Meeting of the Association celebrating a century of service to the independent retail building material dealers of Pennsylvania, Southern New Jersey, Delaware, Maryland, and Washington, D.C. Remembering the past... Eastern was founded in 1892 and is among the oldest associations in the country. Honoring the present... Eastern's combined membership numbers more than 750 firms. A professional staff of 24, coordinate three major spheres of activity: legislative and regulatory representation, education and training, and combined purchasing power to cut day-to-day operation costs for independent businesses. Looking forward to the future... Eastern is celebrating the start of a new century of service and success. See you at the show!

1992 Eastern Market Flyer

Pennsylvania Lumbermen's Mutual Insurance Company

It was a fact. Insurance companies believed all lumber properties were potential fire hazards. Insurance rates for lumber properties were higher than other enterprises—despite the records that clearly showed an exceptionally low fire loss ratio.

In 1895 a group of prominent eastern lumbermen decided they were fed up with the discriminatory practices of the insurance business. For two years their many protests against the unjust rate structure had not only failed to change the position of the insurance associations, they had backfired.

As an example, Philadelphia lumberman Edward F. Henson had asked the Fire Underwriters' Association of Philadelphia for a lower rate, submitting facts and figures to substantiate his claim. Within days he received a reply: His rate was increased!

With Henson leading the way, the lumbermen organized a mutual fire insurance for lumber operations only. Thus, the Pennsylvania Lumbermen's Mutual Fire Insurance Company was born.

From its headquarters in downtown Philadelphia, PLM provides insurance policies to close to 4,000 lumber businesses. Until recently the firm provided only property insurance. However, in 1981, casualty insurance was added to expand PLM's services to lumber businesses throughout New England, the Southeast, and the Midwest.

The first PLM office was set up in a rented room on the second floor of Henson's lumberyard office at 921 North Delaware Avenue, on the waterfront near Poplar Street. That proved to be an excellent location because most of the Philadelphia lumberyards were located on Delaware Avenue.

The company grew larger and stronger as the years went by. And, although PLM was organized to insure only lumber property, in 1930 the decision was made to write what became known as general business insurance.

Lumber businesses still represent the overwhelming majority of policyholders at PLM. Although the firm's exclusiveness to that market has over time been challenged by its competitors, PLM retains much of the original market that it has held for generations.

To continue to meet the needs of its clients, PLM has, over the past few years, broadened both its sales and marketing efforts, added a Loss Control Division, streamlined its processing, and upgraded its daily servicing capacity. Today the company employs 150 people and taps into a network of approximately 500 lumber insurance brokers to present and sell its policies.

As it heads toward its 100th anniversary, Pennsylvania Lumbermen's Mutual Insurance Company likes to note that it is one of only five insurance companies nationwide that continues to specialize in the lumber business. And the company slogan fits the spirit of growth that has been the benchmark of this firm since 1895: "From a small acorn, a mighty oak."

Reprint from an 1897 newspaper article showing PLM's first office and staff.

PLM fieldmen, circa 1900, at Philadelphia conference.

Cramer's HOME CENTERS

SERVING THE POCONOS FOR OVER 75 YEARS!

Russell C. Cramer (Founder)

Founded in 1915, Cramer's quickly established a reputation for offering quality building materials. Cramer's was an early proponent of storing all its materials in sheds protected from the effects of sun, rain and snow. This proved to become an important advantage over its competition. During those early years Cramer's business was largely with local builders, Pocono area resorts and camps.

East Stroudsburg store, circa 1930.

Cramer's Management Team 1965. First Row (left to right) are Dale Parker, Russell Cramer II, Russell Cramer (Founder), and Clifford Cramer. Second Row (left to right) are Spencer Cramer, Olin Cramer, and Loring Cramer.

Cramer's long history has spanned these important housing industry events: The Great Depression, postwar housing crises, vacation home boom, and the do-it-yourselfer revolution. Cramer's success over the years has been its ability to adapt to ever-changing market needs while maintaining a strong commitment to quality and service. Cramer's currently operates six home center locations in Northeastern Pennsylvania.

Corporate Offices, circa 1991.

Pocono Summit store, circa 1991.

EAST STROUDSBURG • POCONO SUMMIT • PORTLAND • WIND GAP • MOSCOW • EASTON

Saul, Ewing, Remick, and Saul

The lawyers of Saul, Ewing, Remick, and Saul salute Eastern Building Material Dealers Association on its One Hundredth Anniversary. We are proud to have been your lawyers for forty-six of those hundred years. As World War II was coming to a close, some of your members found themselves under government attack for alleged violation of OPA regulations. The association turned to Walter Biddle Saul who, with Fred VanDenbergh (newly returned from Army service), brought the matter to a successful conclusion.

Since that first assignment, our lawyers have worked with the association on many projects. In 1948, we helped you to establish the Group Insurance Trust and, a few years later, the Retirement Program. These continue growing all the time. We have handled your occasional lawsuits, have given advice on all sorts of questions, and have participated in your seminars and training programs.

In the early days, Fred VanDenbergh was in charge of the firm's relationship with the association. Lowell Thomas succeeded him. Both of them enjoyed this work and recall with affection the many friends they made among the members and staff. They will always remember their friendships with deceased association executives Bob Jones, Russ Allen, and Gerald Anderson. We look forward to continuing our work with Dave Kreidler and Harry Johnson.

Saul Ewing had eighteen lawyers when you first came to us. We have over 160 now in Philadelphia; Washington; New York; Wilmington; Voorhees, New Jersey; and Great Valley, Pennsylvania. Together we wish you the best of luck in your second century!

Frederick A. VanDenbergh, Jr., and Lowell S. Thomas, Jr.

Morgan Distribution

The Morgan Distribution organization traces its origins back to 1855 when two Welsh immigrants, Richard and John Morgan, purchased a small planning mill in Oshkosh, Wisconsin. In spite of a succession of fires which destroyed their early manufacturing facilities, the company survived and prospered. A portion of the present factory and office were built in 1896 and 1897 after the last fire. From these humble beginnings Morgan Manufacturing has grown to a plant comprising some 750,000 square feet of manufacturing space on 26 acres of land.

In 1911, Morgan entered the wholesale millwork distribution business in the East with the acquisition of the Baltimore Sash and Door Company which was renamed Morgan Millwork Company. At about the same time, another distribution facility, Morgan Sash and Door Company, was established in Chicago.

During the 1920s, branches were established at Greenville, South Carolina; Jersey City, New Jersey; and a sales office was opened in New Haven, Connecticut. Early in the Depression the decision was made to distribute Morgan manufactured products through independent wholesale distributors in the Northeast and so Jersey City and New Haven were closed. Greenville was also closed at this time. In 1932 a branch office was opened in Wilmington, Delaware.

In 1954, the Decatur, Illinois warehouse was established as a branch of Chicago. Morgan Millwork Company acquired the Radford and Sanders Company of Baltimore and Harrisburg, Pennsylvania, in 1954. Radford's Baltimore operation was closed, but Harrisburg was retained. At the same time, a new warehouse was established in Arlington, Virginia, the forerunner of the Alexandria branch which, in 1991, was relocated to Gainesville, Virginia. In 1965, the Portsmouth, Virginia branch was opened and relocated to Chesapeake in 1984.

In 1956, the Harrisburg warehouse was relocated to Camp Hill, Pennsylvania, and in 1976, moved to the present facility in Mechanicsburg, Pennsylvania.

Following the acquisition of Morgan by Combustion Engineering in late 1972, the existing warehouses of C-E Building Products, which distributed aluminum products to Florida builders, were expanded to include millwork products. However, the attempt to combine aluminum and wood products and to convert from one-step to two-step distribution was unsuccessful and these locations gradually were closed.

In the meantime, Flint Sash and Door Company of Flint and Saginaw, Michigan, was acquired in late 1973. In 1978 these warehouses were consolidated into a single modern facility at Birch Run, Michigan.

During 1975, a new warehouse was established in Columbia, South Carolina. This operation moved to a new larger modern facility in 1981. Also in 1975 the Chicago facility was closed and Baltimore was closed in 1978, though all of the Baltimore trading area was retained and serviced by Alexandria, Mechanicsburg, and Newark.

Early in 1979, Rust Sash and Door Company of Lenexa, Kansas, was acquired. Morgan Distribution was further expanded in 1982 when Robbins Door and Sash Company, with eight locations in New York, Pennsylvania, and Maryland, decided to liquidate their business. As a result of this, Morgan Distribution center was established at Scranton, Pennsylvania.

In January 1984, Combustion Engineering sold the Morgan organization to a group of private investors headed by Chairman of the Board Frank Hawley. The new company was named Morgan Products Limited and became a public company when a public stock offering was made in October 1985. The most recent expansion of Morgan Distribution was in June 1987, with the opening of a distribution center in West Chicago, Illinois, to service the northern Illinois market.

With ten warehouse locations and over seven hundred employees, Morgan Distribution is today one of the largest wholesale millwork distributors in the United States. Yet each of these locations has successfully maintained its individual character as a regional supplier of quality brand-name millwork products.

Distribution Center locations currently serving the EBMDA trading area are: Mechanicsburg and Scranton, Pennsylvania; Newark, Delaware; and Washington, D.C.

Some of the nationally known brands which are distributed are Anderson Windowwalls, Morgan Doors and Woodwork, Pease and Therma-Tru Metal and Fiberglass Door Systems, Armstrong Ceiling Systems, Webb Decorative Windows and Louvers, Stephenson Cupolas and Cellwood Shutters, along with a number of well-known brands of specialty products.

Russell Plywood, Inc.

Russell Plywood Inc., Wholesale Distributors of Reading, Pennsylvania, officially began business on September 1, 1986. However, the company's history goes back to 1946 when three DiGiallorenzos—father and two brothers—started Plywood Distributors at Richmond and York streets in Philadelphia. With the retirement of the father, John DiGiallorenzo, in 1953, one son, Russell, and a partner, established Penn-Valley Plywood, Inc. The company grew and conducted business until September 1986, when it split and the Reading branch, which had begun operation in 1968, was taken over by Russell DiGiallorenzo and named Russell Plywood.

Having begun with two employees, Russell Plywood now has thirty-six, many of whom have a long tenure with the company. Chief among these is manager Darryl Milkins, who has been with the company for twenty-two years.

Since 1986, Russell Plywood, Inc., has continued to grow each year. In 1988 they doubled their warehouse space and have more than doubled their office area, thanks to their loyal customers, friends, and employees, In 1991 a branch was opened in New Castle, Delaware, to service Eastern Pennsylvania, New Jersey, Maryland, the District of Columbia and Northern Virginia. It has been their policy to act, say, and do with the highest integrity, quality, and performance. The company remains a family business with both Russell Sr. and Jr. actively involved in the affairs of the company.

Russell Plywood is proud to look back over its history to a fine customer relationship, the development of major product innovations to suit the marketplace, and to look to the future to continued growth as they continue their efforts to advance the industry as a whole. "We realize that for any industry to keep abreast of the ever changing tomes, we should continuously strive to work in the interest of the general public to promote the free enterprise system of the United States. To this we dedicate ourselves, united in the belief in God, to the enlightenment of the human spirit and to enrich and broaden the pattern of the American way of life."

RUSSELL PLYWOOD, INC.
WHOLESALE DISTRIBUTORS
401 OLD WYOMISSING ROAD • READING, PA 19611-1507 • 215/374-3206

Willis Corroon

Although a new corporate entity, Willis Corroon Corporation of Pennsylvania can trace its roots to organizations on both sides of the Atlantic and to three companies that trace their histories to the early nineteenth century. The shares of Willis Corroon are listed on the London Stock Exchange—where the company is included on the index of the one hundred largest companies in terms of market capitalization—and on the New York and Pacific exchanges.

Noyes Services began in 1885 in Swarthmore, Pennsylvania. It served primarily as a real estate firm until the 1950s, when the Insurance Company of North America developed the Homeowner's Policy and the insurance business quickly outpaced real estate. In 1989, Noyes Services joined Corroon & Black, the fourth largest publicly owned United States insurance brokerage firm. Corroon & Black's foundations were laid in 1905 with the formation in New York of R. A. Corroon & Company; in 1929 it became the first publicly held insurance broker.

Willis Faber can trace its origins to early nineteenth century England and the formation of a number of Lloyd's banking and insurance agency firms. In October 1990, Willis Faber and Corroon & Black merged to create Willis Corroon. Today, under the name Willis Corroon Group plc, the Group functions as a global entity, with service lines responding to client and regional requirements. The new structure will streamline the Group's operations and focus its services to clients on a worldwide basis.

Beginning with Noyes Services, Willis Corroon has provided expert services to the Eastern Building Material Dealers Association and its member dealers on a wide range of issues involving insurance and investment needs. Given our combined histories, we look forward to that continued close service over the next hundred years.

WILLIS CORROON

William H. Fritz Lumber

William H. Fritz, Inc., of Berwyn, Pennsylvania, is a fifth-generation family-owned lumber business. The exact founding date of the company is unknown, but there are references to Fritz's lumberyard as early as 1863, since it is recorded that he sold lumber to the Leopard and Ogden schools in 1863 and 1864.

Succeeding to the business after the death of his father, William H. Fritz went on to become one of Berwyn's leading businessmen. He served as president of the Berwyn Bank and a leader in the Trinity Presbyterian Church. William H. Fritz, Jr., took over the management of the company in 1941 upon the death of his father, and William H. Fritz III entered the business in 1956.

Today, the fifth generation of the Fritz family, Howard and Andrew, are active in the business, and the family's involvement in the community continues, although not like it was earlier in the century when the offices of the Easttown Township government were in the company's quarters. However, the family commitment to the community continues as does its commitment to quality service to the customer.

W. H. Fritz coal loader, circa 1900.

I. F. March Sons

Near the turn of the century, Isaac F. March moved his lumber and feed business from Monocasy to Bridgeport, Pennsylvania, with his son Mathias. I. F. March Sons was a founding and charter member of the lumber association.

After his father's retirement, the company grew quite rapidly under the direction of Mathias March. The company's adaptation to mechanization and machinery efficiently met the growing demands of an expanding Montgomery County area.

In the early 1900s, new operations were added to the lumber business. These included a planing mill and an extensive wooden box factory. The company then, and to this day, produces wooden boxes and shipping crates for packaging worldwide industrial shipments. The 1920s marked another significant change in the company's operation, when truck transportation replaced horses.

Mathias March led the company through the Great Depression of the 1930s by building homes in the surrounding area. This provided the lumber business, its employees and its customer base with consistent work. This practice is still followed today by I. F. March Sons during the currently difficult economy.

During the 1940s and 1950s the company was led by Mathias' sons William and George, Sr. This was a time of hard work coupled with dramatic changes in transportation and distribution. Changes in product lines, such as the inclusion of plywood and drywall, provided a challenge to the fast-paced post war-driven economy. In 1956, a fire nearly engulfed the entire lumberyard.

During the 1960s and 1970s, George K. March, Jr., took over the role of leadership of the firm. Under his direction, the company acquired W. H. Kneas Lumber Company, the firm's closest and largest competitor, which enabled the company to expand in a changing marketplace. In 1974, another tragic fire destroyed much of the W. H. Kneas Lumber facility. The company rebuilt and combined operations between the Bridgeport and Norristown facilities with a new efficient base of operations.

Under the direction of David W. March, the company has expanded to three locations in Montgomery County, Pennsylvania, supplying primarily the builder/contractor. The company has also expanded its operation into residential real estate development. Currently, the company has over one thousand home sites available to its builder customers. David W. March believes that vertical integration is the key to future growth and longevity.

"Although we are proud of the five generations of our family business, we believe that, ultimately, it is our employees, both past and present, that have allowed I. F. March Sons to enjoy its past and prosperity."

Index

A

Associate Members, 163
Association Offices, 151–153
Advisory Board, 78
Allegheny County, 45
Allen, Russell J., 94, 96, 98, 99, 125, 127, 150
American Society of Association Executives (ASAE), 12, 99, 114
Annual Management Conference, 111
Annual Retail Lumber Dealers Short Course, 92
Antitrust Division, 98
Arnold, Thurman, 79
Associate Member Committee, 107, 172
Association Credit Bureau, 63
Atlantic Deeper Waterways Association, 58
Atticks and Britcher, 102, 144
Auld, John H., 102

B

Baltimore Sash and Door Company, 180
Barnhart log loader, 21
Baumgardner, H. K., 42, 51
Beitzel, Jacob, 42
Benetez, Carol, 174
Bennett, Mary, 174
Blue Eagle, 72
Board of Directors, 42, 43, 46, 47, 48, 49, 51, 54, 55, 56, 58, 61, 74, 75, 77, 78, 79, 80, 85, 90, 91, 92, 93, 95, 96, 97, 99, 100, 101, 120, 125, 131, 132, 133, 136, 137, 138, 139, 140
Boston Agreement, 47
Boughner, Clara, 174
Bowers, G. Hunter (Hunt), 123, 124
Bradley, Joseph C., Jr., 112, 173
Brady, Mary Ann, 174
Brandow, O. M., 46, 47, 48, 56
Brosius and Smedley, 75
Brosius, Joseph W., 90, 124
Brosius-Eliason, Inc., 75, 76, 147
Brown, Richard W., 99, 173

Buckley, James, 74
Building Committee, 97, 107
Building Material Dealers Association of Eastern Pennsylvania, 64
Building Material Retailer, 100, 103
Bulletin Building, 151
Bunting, Betty, 92
Bureau of Information, 48
Bures, Ray, 125
Business Development Committee, 107, 168
Business Forms Program, 96
Butterick Publishing Company, 58
Bylaws Committee, 106, 165

C

C-E Building Products, 181
Centennial Companies, 154
Centennial Committee, 9, 170
Century Club, 154
Certified Lumber Standards, 81
Charron, Dave, 125
Chupein, Dennis, 174
Clark, Robert E., 173
Code of Ethical Standards, 100, 101
Combustion Engineering, 181
Commissioner of Internal Revenue, 75
Committee on Arbitration, 57
Committee on Complaints, 47
Committee on Constitution and Bylaws, 42
Committee on Enlargement of Organization, 47
Committees on Entertainment, 52, 119
Committee on Increasing Membership, 44
Committee on Insurance, 47
Committee on Legislation, 44, 46, 52
Committee on Permanent Organization, 42
Committee on Transportation, 44, 47
"Compensatory Pricing of Sheet Materials—Cut to Size," 98
"Constitutiion and By-Laws," 45, 46, 47, 48, 52, 53, 54, 55, 56
Convention Advisory Committee, 107, 171
Cook, Robert E., and Son, 149
Coolidge, Calvin, 123
Corroon & Black, 115, 116, 183
Corroon, R. A., and Company, 183
Costello, Bill, 123
Craig, J. W., 45, 49
Cramer, Clifford, 178
Cramer, Loring, 178
Cramer, Olin, 178
Cramer, R. C., Lumber Company, 132
Cramer, Russell C., 178
Cramer, Russell, II, 178

Cramer, Spencer, 178
Cramer's Home Centers, 178

D

Dealers Directory and Buyers Guide, The, 83, 96, 139, 140
Delaware Lumber Dealers for Good Government, Inc., 110
Department of Justice, 79
Depression, 19, 71
DiGiallorenzo, John, 182
DiGiallorenzo, Russell, 182
DiGiallorenzo, Russell, Jr., 182
Dillingham, Linc, 125
Doctrine of "reasonable restraint," 17
Dodson Insurance Group, 115

E

Earle, Governor, 79
Eastern Building Material Dealers Association (EBMDA), 20, 100, 101, 108, 110, 112, 114, 116, 117, 133
Eastern Building Material Dealers Education Foundation, 100, 107
Eastern Group Trust, 101, 107, 112, 113
Eastern Lumber Salesmen Association, 56
Eastern Market, 101, 125
Eastern Pennsylvania Lumber Dealers Association, 61
Eastern Retirement Trust, 107, 114, 115
Eastern States Retail Lumber Dealers Association, 20, 47
Edgecomb, I. M., and Sons, 28
Education Committee, 107, 167
Eisenhauer, McLea and Company, 61, 66
Elected Leadership, 155
Elverson Supply, 149
Emporium Lumber Company, 30
Executive Committee, 43, 44, 46, 47, 52, 56, 78, 85, 93, 120, 121, 138, 140

F

Fair Labor Standards Act, 83
Federal Housing Administration (FHA), 90
Fehrenbach, A. J., 138, 139
Ferrara, Nicole, 174
Ferretti, Bruce, 10
Finance Committee, 81
Finley, Nancy, 174
First Pennsylvania Bank, 114
First Pennsylvania Bank Building, 91, 153
First World War, 18, 19
Fleck, E. Thomas, 98, 114, 173
Fleming, Robert, 123
Flint Sash and Door Company, 181
Frisbee Lumber Company, 146
Fritz, Andrew, 184

Fritz, Howard, 184
Fritz, William H., 184
Fritz, William H., Jr., 184
Fritz, William H., Lumber, 184
Fritz, William H., III, 184

G

Galliher & Huguely, 83
Garden State Lumber Dealers for Good Politics, Inc., 110
Gavel Cavaliers, 157
Girard Trust Building, 81, 152
Golden Anniversary Convention, 118, 119
Goodling, William, 102
Goodyear Lumber Company, 21
Graff, Charles, 91, 94, 123
Grandizio, Anna A., 173
Green, H. G., Lumber and Coal Company, 69
Green, Jonas, Jr., 173
Group Insurance Trust, 105
Grover, William, 99

H

Haenn, Joseph E., Jr., 89, 94, 132
Hamilton, Mary, 174
Haney, C., 42
Hanyen, Frederick C., 49, 52
Harman, Edgar B., 99, 125
Hawley, Frank, 181
Hazard, J. F., 50
Heller, F. P., 42
Henson, Edward F., 177
Hess, Dr., 138

Hood, Art, 123
Hooper, Nancy, 174
Hoover, Herbert, 18, 122
Howarth, J. W., 47

I

Illinois Lumber and Builders' Supply Dealers' Association, 20
Inspection Service, 74
Instant Lumber Piece Price, 89
Insurance, 129–133
Integrity Bank Building, 152
Interstate Commerce Commisssion (ICC), 52, 53
Iowa Retail Lumber Dealers' Association, 19

J

James, W. M., 45, 50
Johns-Manville Corporation, 122, 124, 125
Johnson, Harry H., III, 13, 94, 96, 99, 112, 114, 115, 125
Johnson, Hugh S., 123
Jones, Robert A. (Bob), 78, 79, 80, 83, 87, 88, 89, 90, 94, 125, 131, 139, 150
Junior Division, 56
Justice Department, 97

K

Kaufman, J. L., 42, 51,
Kauffman, Richard A., 102
Kavanaugh, Marietta, 174
Keating Summit, Pennsylvania, 27, 30
Keck, S. H., 45, 47, 49
Keith, D. J., 136
Keller, Charles B., 46
Kildoo, Bill, 96
Kneas, W. H., Lumber Company, 185
Kreidler, David B., 13, 98, 99, 151
Krimm, Charles, 91

L

Lackawanna County, 50
Lancaster, George F., 39
Lance, George F., 42, 45, 47, 49
Latshaw, Ray, 80, 131, 139
Laudig, B. F., 48, 150
Lee, G. F., 47
Legislative Committee, 106, 166
Lehigh Lumber Company, 91, 146
Linn, Linda, 174
Little Kettle Creek, 24
Little Pine Creek, 25
Logging, 17, 18, 21–26, 63
Loizeaux J. D., Lumber, 67

Loysville, Pennsylvania, 31
Luck, J. Howard, 11
Ludwig, Fred A., 123
Lumber Secretaries' Association, 20
Lumber Short Course, 101
Lumbering, 15–16, 27–31, 61, 62, 64, 69
Lumberman's Credit Association, 52
Lumberman's Exchange, 45
Lumbermen's Merchandising Corporation, 74
Lumbermen's Purchasing Corporation (LPC), 74

M

M.A.L.A. Group Insurance Trust, 96, 98
M.A.L.A. Retirement Benefit Programs, 98
M.A.L.A. Retirement Trust, 96, 100
MALA. See Middle Atlantic Lumbermens Association
MALA, Inc., 77, 95, 96, 98, 100, 107, 109, 132, 133
Malone, Pat, 124
Malone, Watson, III, 85
Management Development Committee, 107
Management Development Program, 101, 110
March, David W., 185
March, George K., Jr., 185
March, George K., Sr., 185
March, I. F., Sons, 185
March, Issac F., 185
March, Mathias, 185
March, William, 185
MARM, Inc. See Mid-Atlantic Risk Management, Inc.
Martin, J. Frederick (Fred), 52, 73, 79, 80, 131, 137, 150
Maryland Lumber Company, 90
Maryland Lumber Dealers for Good Government, Inc., 110
Masten, Pennsylvania, 28
McIllvain, J. Gibson, and Company, 62
McEvoy, John F., 122
McEwing, Bill, 140
Member Services department, 105, 116, 117
Members, Active Dealer, 158–162
Members, Associate, 163–164
Members, Founding, 40
Membership Bulletin, The, 110
Membership Committee, 56
Membership Development Committee, 107, 169
Membership Development Program Committee, 169
Mid-Atlantic Risk Management, Inc. (MARM), 99, 101, 106, 107, 115, 116, 133
Middle Atlantic Lumbermens Association (MALA), 71–104, 105, 106, 107, 109, 124, 132
Middle Atlantic Lumbermens Association Group Insurance, 87, 131
Middle Atlantic Service Corporation, 96
Milkins, Darryl, 182

Mitchell, John D., 98, 99, 100
Mizell Lumber and Hardware Company, Inc., 58, 131
Moore, Bob, 125
Moore Business Forms, 117
Morgan Distribution, 180
Morgan Millwork Company, 180
Morgan, John, 180
Morgan Products Limited, 181
Morgan, Richard, 180
Morgan Sash and Door Company, 180
Moyer and Shearer, 33
Moyer, E. K., 58

N

Name Committe, 92
National Association of Manufacturers, 52, 72
National Lumber and Building Materials Dealers Association (NLBMDA), 11, 20, 132
National Recovery Administration (NRA), 19, 72, 123
National Retail Lumber Dealers Association (NRLDA), 63, 81, 83, 85, 87, 92
National Wholesale Lumber Dealers Association, 47
National Wholesalers Association, 44, 45, 47
Nelco Lumber and Home Center, 77
New Deal, 19, 71
New Jersey Lumber and Building Material Dealers Association, 96
New York Journal, 46
New York Lumber Trade Journal, 44, 135
Newcomb, Michelle, 174
NLBMDA, 11, 20, 132
Nominating Committee, 106, 165

Norfolk, Va., Lumber Exchange, 31
Norman, William E., 93
Northern New Jersey Lumbermen's Association, 91
Noyes Services, 116, 183

O

O'Connor, Ada, 174
Ohio Valley Lumber Company, 65, 76
"Opinionaire," 93
O'Reilly, James A., 39
Oswald, Kermit, 139
Otis Building, 151
Owens Corning Fiberglas, 94

P

Parker, Dale, 178
Passmore Supply, 148
Patterson, J. E., 47
Penn State University, 111
Penn-Valley Plywood, Inc., 182
Penn, William, 15
Pennsylvania Lumber Dealers for Good Government, Inc., 110
Pennsylvania Lumber Museum, 15, 17, 18, 21, 22, 23, 24, 25, 26, 27, 28, 29, 30, 31, 38
Pennsylvania Lumberman's Association, 49, 50, 57
Pennsylvania Lumbermen's Association (PLA), 54, 58, 63, 64
Pennsylvania Lumbermen's Mutual Fire Insurance Company, 45, 63, 80, 129, 130, 177
Pennsylvania Lumbermen's Mutual Insurance Company, 115, 116, 177
Pennsylvania Lumbermen's Protective Association (PLPA), 16, 39, 41, 42, 43, 44, 46, 47, 48, 49, 135, 136
Pennsylvania State College, 85
Pennsylvania State Forestry Association, 53
Pension Plan Committee, 87
Philadelphia Lumber Exchange, 31, 45, 63, 120, 129
Philadelphia Lumbermen's Exchange, 91
Plan, The, 55, 56, 60, 63, 66, 72, 75, 78, 79, 82, 83, 86, 87, 88, 89, 100, 134, 136, 137, 138, 139, 140
Political Action Committees, 110
Potter County, Pennsylvania, 24
Presidents, 156
Price fixing, 16
Property/Casualty Program, 101, 107
Publications, 135–141
Pyle's Home and Supply, 148

R

Radford and Sanders Company, 180
Randenbush, George W., 39
Rearick, J. H., and Son, Inc., 102, 144

Rearick, John H., 102
Rearick, Mary K., 102
"Red Book," 52
Reiter, J. H., 63, 137, 138
Retail Lumber Dealers Association of Maryland, 91
Retail Lumber Dealers' Association of Pittsburgh, 54
Retail Lumber Institute, 85, 94
Retail Lumber Secretaries Association, 20
Retail Lumber Trade Associations, 19–20
Retirement Programs Trust, 105
Risser, Jay F., 113
Robbins Door and Sash Company, 181
Rockefeller, John D., 123
Russell Plywood, 182
Rust Sash and Door Company, 181
Ryman, Leslie S., 47

S

Saul, Ewing, Remick and Harrison, 85
Saul, Ewing, Remick and Saul, 95, 131, 179
Saul, Walter Biddle, 179
Scholtz, Edwin F., 96, 99
Searce, H. C., 20
Sears-Roebuck, 59, 74
Secretary's Association, 47
Sener, W. Z., 45, 47, 50
Shearer, A. K., 34, 35, 36
Shearer, A. K., Company, 31, 32, 33, 34, 35, 36, 37
Sheldon School of Sales Training, 85
Sherman Act of 1890, 17
Shield Brothers, 26
Smick and Harris, 57
Smick, B. Harold, Jr., 7, 8, 102
Smick Lumber and Building Materials Center, 57, 128, 132, 145
Smith, Adam, 16
Smith, Walter, 64
Snowden, T. J., 42, 46, 47, 48, 56
"Splinters," 140
Stambaugh, Eileen, 174
Store Fixtures Program, 96
Strategic Planning Committee, 107
Straus, S. J., 43
Sturdevant, S. H., 42, 46, 47, 48, 56
Successful Merchant, The, 138
Supreme Court of the United States, 17, 19
Susquehanna River, West Branch, 25
Sweden Valley, 27

T

Tatem, Joseph W., 136

Taylor, R. William, 12
"10 Point Program," 85
Thomas, Lowell S., Jr., 95, 179
Thompson, Albert J., 57
Thompson and Baird, Esq, 79
Tinsman Brothers, 58, 65
Top Management Workshop, 85, 93, 101
Trade Associations, 16–19
Trexler, Harry C., 42, 51
2 Penn Center Plaza, 152

U

U.S. Department of Agriculture, 53
United Association of Lumbermen, 20
United Lumberman's Association, 45
United States v. Eastern States Retail Lumber Dealers Association, 17, 57

V

VanDenbergh, Frederick A., 95, 179
Virginia Lumbermen's Association, 91, 96

W

Wallace, Henry, 123
Walton, John B., 112
War Industries Board, 18
West Virginia Group Insurance Trust, 96
West Virginia Lumbermens Association, 91, 96, 132
Western Pennsylvania Lumber Dealers Association, 91, 96
Western Pennsylvania Retailers Association, 60
Willard, E. M., 45, 47
Willis Corroon Group plc, 183
Willis Faber, 183
Wolf, Bill, 124
Wood, Bob, 125

Z

Zook, Gary W., 173

About the Author

Dr. George W. Franz
Photo courtesy of Penn State Delaware County Campus

George W. Franz holds an A.B. degree from Muhlenberg College and an M.A. and Ph.D. from Rutgers University. He has been teaching history at Penn State Delaware County Campus for twenty-two years, where he is associate professor of history and American studies and also serves as honors coordinator. Dr. Franz has won the Outstanding Teacher Award at the campus and the University's George W. Atherton Award for Excellence in Undergraduate Education.

Dr. Franz is the author of *Paxton: A Study of Community Structure and Mobility in the Colonial Pennsylvania Backcountry* (Garland, 1989) and he was the project director and editor of *The Papers of Martin Van Buren*, a microfilm edition of the papers of the eighth president of the United States. His reviews have appeared in *Pennsylvania Magazine of History and Biography, Pennsylvania History* and *The Journal of American History*.

He lives in Chadds Ford, Pennsylvania, with his wife, Kammy, and his two children, David and Wendy. In the community, he is active in Boy Scouting, serving as secretary/treasurer of Troop 31, and is president of the Chadds Ford Historical Society. In addition, he is a gubernatorial appointee to the Brandywine Battlefield Commission.

Pennsylvania Lumbermens' Annual Dinner,
December 1, 1925.
Photo courtesy of EBMDA